REVERSIBILITY OF ACIDIFICATION

Proceedings of a workshop jointly organised by the Royal Norwegian Council for Scientific and Industrial Research (NTNF) and the Commission of the European Communities, Brussels, Belgium, at Grimstad, Norway, 9–11 June 1986.

This workshop was organised as part of concerted action (COST 612) on effects of air pollution on terrestrial and aquatic ecosystems (working group II: effects of air pollution on aquatic ecosystems).

This is Report No. 8 in the Air Pollution Report Series of the Environmental Research Programme of the Commission of the European Communities, Directorate-General for Science, Research and Development. For more information concerning this series, please contact:

Mr H. Ott
CEC—DG XII/G1
200 rue de la Loi
B-1040 Brussels
Belgium

REVERSIBILITY OF ACIDIFICATION

Edited by

H. BARTH

Commission of the European Communities, Brussels, Belgium

ELSEVIER APPLIED SCIENCE
LONDON and NEW YORK

ELSEVIER APPLIED SCIENCE PUBLISHERS LTD
Crown House, Linton Road, Barking, Essex IG11 8JU, England

Sole Distributor in the USA and Canada
ELSEVIER SCIENCE PUBLISHING CO., INC.
52 Vanderbilt Avenue, New York, NY 10017, USA

WITH 31 TABLES AND 73 ILLUSTRATIONS

© ECSC, EEC, EAEC, BRUSSELS AND LUXEMBOURG, 1987

British Library Cataloguing in Publication Data

Reversibility of acidification.
    1. Freshwater ecology   2. Acid rain—Environmental aspects
    I. Barth, H.
    574.5'2632   QH541.5.F7

ISBN 1-85166-172-7

Library of Congress CIP data applied for

Publication arrangements by Commission of the European Communities, Directorate-General Telecommunications, Information Industries and Innovation, Luxembourg

EUR 11287

LEGAL NOTICE
Neither the Commission of the European Communities nor any person acting on behalf of the Commission is responsible for the use which might be made of the following information.

Printed in Great Britain by Galliard (Printers) Ltd, Great Yarmouth

PREFACE

The workshop reported in this volume reviewed knowledge of the effects on aquatic ecosystems of changes in atmospheric deposition of acidifying components, in particular, to examine the question of whether aquatic ecosystems will recover from acidification once acidifying depositions are reduced. Emphasis is given to modelling of the expected effects of reduced atmospheric sulphur and nitrogen deposition. Land-use changes as they relate to the effects of changes in deposition are also taken into account, but mitigation by measures such as liming was excluded from the topics covered by the workshop.

The papers include both theoretical and experimental research results. The workshop was held under the auspices of the COST programme (European Cooperation in Science and Technology), one of the main aims of the workshop being to identify possible fruitful areas of cooperation within this research topic which could be pursued within the COST framework.

C O N T E N T S

SESSION I - FIELD STUDIES AND EXPERIMENTS

Recovery of canadian lakes from acidification

Rain-project : results after 2 years of treatment

Some aspects of the chemical speciation of aluminium in acid surface waters

Exposure of small-scale aquatic systems to various deposition levels of ammonium, sulphate and acid rain

The chemical and biological features of poorly buffered irish lakes

Acidification problems of freshwaters: trophic relationships

Physiological study on the recovery of rainbow trout (Salmo Gairdneri Richardson) from acid and al stress

The development of the acid lake gribsø in Denmark after 1950

Chemistry of atmospheric deposition and lake acidification in Nothern Italy, with emphasis on the role of ammonia

Buffering mechanisms in acidified alpine lakes

Acidity mitigation in a small upland lake

RECOVERY OF CANADIAN LAKES FROM ACIDIFICATION

D.W. Schindler
Department of Fisheries and Oceans
Freshwater Institute
501 University Crescent
Winnipeg, MB   R3T 2N6   Canada

Summary

The recovery of Canadian lakes from acidification has been studied in the Sudbury area, where emissions and deposition of sulfur have decreased since the mid 1970's, and in Lake 223 of the Experimental Lakes Area, which was experimentally acidified to pH 5 between 1976 and 1983.

The concentrations of sulfate decreased rapidly in all lakes after sulfur inputs were reduced. The pH values increased by from 0.4 to 1.7 units, and increases in alkalinity were observed in lakes where measurements were made. The increases in alkalinity appeared to be caused by the much greater rate of decrease in sulfate than in base cations. Recoveries were equal to, or slightly slower than, predicted from water renewal rates.

In Lake 223, reproduction of all species of fishes had ceased when the lake was acidified below pH 5.4. When the pH was raised to this value during the recovery of the lake, reproduction resumed in all species which remained, with the exception of lake trout. The condition of lake trout improved when the pH was increased, but so far they have not reproduced. Animal species eliminated by acidification, including Mysis relicta, Orconectes virilis, Cottus cognatus and Pimephales promelas, have not returned to the lake so far. A large population of Culaea inconstans seems to have partially taken over the niches vacated by these species. Phytoplankton diversity increased as the pH increased, and algal species eliminated by acidification returned to the lake.

1. INTRODUCTION

Until very recently, it was thought that acidified lakes might be incapable of recovering without liming or some equivalent addition of alkalinity-generating chemicals. The buffering capacity of natural lakes was thought to derive from geochemical weathering processes in their terrestrial watersheds, and it was believed that once these sources were exhausted, they were gone forever. As a result, there has been some dispute over whether reducing acidifying emissions would cause any beneficial effects on lakes.

Recently, in situ biological processes (reduction of nitrate and

sulfate) have been shown to be major, and in some cases dominant
buffering mechanisms in the lakes of acid-vulnerable areas (reviewed
by Schindler 1986). These processes are unimpeded by acidification,
operating at pH values of 4 or lower (Bayley et al. 1986). In fact,
they usually increase as the concentrations of biologically-reactive
strong acid anions (SO_4^{2-}, NO_3^-) increase (Cook et al. 1986;
Schindler 1985; Schindler et al. 1986). As a result, the latter
authors predicted that lakes would recover rapidly once the influx of
acidifying substances to lakes was decreased. This paper summarized
three Canadian studies, of the responses of eight lakes to decreased
input of sulfate and SO_2.

2. STUDY AREAS

The Sudbury, Ontario area is notorious for massive emissions of
SO_2 since the late 19th century, causing widespread and rapid
acidification of lakes (Gorham and Gordon 1960; Conroy et al. 1974;
Beamish and Harvey 1972). The Sudbury area is one of the few places
on earth where emissions of sulfur oxides and deposition of SO_4^{2-}
have decreased. In 1972, the Coniston smelter, which emitted 200,000
tonnes of SO_2 per year, was closed. In the same year, the Copper
Cliff smelter replaced three smokestacks of <200 m in height with a
single stack of 381 m, causing a further local decrease in acidifying
deposition. Overall, the emissions of SO_2 at Sudbury decreased from
an average of 2.2 x 10^6 tonnes in 1950-1972 to 0.65 tonnes in
1979-1983, a reduction of 70% (Fig. 1). Corresponding reductions in
bulk deposition from 1970 to 1977 were about 75% (Hutchinson and Havas
1986).

Baby Lake and Alice Lake are approximately 1 km southwest of the
Coniston smelter. The basin of Baby Lake has been denuded of
vegetation and organic soils by high levels of SO_2, wood cutting to
feed the Coniston roastbed, and high doses of nickel, copper and other
toxic trace metals (Hutchinson and Whitby 1977). Alice Lake receives
drainage from Baby Lake, and has a drainage basin covered by deep
glacial till. Typha latifolia and Phragmites communis still grow at
the lake margin, while grasses Deschampsia cespitosa and some trees
(Quercus rubra, Acer rubrum, Populus tremuloides) still grow in the
catchment. Rapid erosion causes the lake to be quite turbid. Water
chemistry data for these two lakes were available from 1968 until
1984, from a variety of sources (Hutchinson and Havas 1986).

Middle, Hannah, Lohi, Clearwater and Swan Lakes are located
slightly farther from sources of SO_2 (Table 1). The first four lakes
were monitored continuously by the Ontario Ministry of Environment
from 1973 onward. The last was sampled twice, in 1977 and 1982.
Middle and Hannah Lakes were manipulated in 1973 by experimental
additions of base ($CaCO_3$ and $Ca(OH)_2$) plus phosphorus, while Lohi Lake
received base alone. The latter lake reacidified to pH 4.8 by 1978,
while the former lakes maintained pH values >6.5. Clearwater and Swan
Lakes were not manipulated in any way (Dillon et al. 1986).

Lake 223 in the Experimental Lakes Area lies in a region of low
acid deposition (wet SO_4^{2-} deposition <5 kg ha^{-1}·yr^{-1}). The
drainage basin is in a pristine state. The lake lies 3.5 km from the
nearest road, and is accessible only by crossing two lakes and two
portages. After two years of "background" study in 1974-1975, direct
experimental additions of sulfuric acid were used to lower the pH of

the lake from natural values of >6.5 to 5.0–5.1 between 1976 and
1981. The pH was kept at 5.0 to 5.1 for 1981–1983. Biological
results of this study were summarized by Schindler et al. (1985), and
chemical results by Schindler and Turner (1982) and Cook et al.
(1986).

Beginning in 1984, sulfuric acid additions were decreased
sufficiently to allow the lake to recover to pH 5.4. In 1985, one of
the wettest years on record, the pH was allowed to increase still
more, to 5.5–5.6.

Morphological features of all of the above lakes are summarized
in Table 1.

3. CHEMICAL RECOVERIES
Sulfate

After emissions and deposition decreased, sulfate concentrations
decreased in all of the lakes studied near Sudbury. The greatest
decreases were in the lakes nearest the sources of emissions (sulfate
in Baby and Alice Lakes decreased by 58 and 62 percent, respectively),
except for Swan Lake, where SO_4^{2-} decreased 62%. It is hypothesized
that the onset of anoxic conditions in the hypolimnion of Swan Lake
during the period of study enhanced removal of sulfate by microbial
reduction and sedimentation (Dillon et al. 1986). Sulfate decreases
in Clearwater, Middle, Hannah and Lohi Lakes ranged from 25 to 44%
(Table 2).

Sulfate in Lake 223 decreased by 13% in two years. While this
figure may seem low, experimental sulfate additions were regulated to
maintain a preselected pH, rather than keeping sulfate input
constant. The high precipitation year in 1985 required large inputs
of sulfuric acid to keep the pH from recovering more rapidly than
planned.

pH and Hydrogen Ion

In all cases where valid comparisons could be made, hydrogen ion
concentration decreased much more rapidly than sulfate. Decreases
ranged from 59% in Clearwater Lake to 98% in Alice Lake (Table 2).
Once again, the greatest recoveries were observed in lakes closest to
the sources of emissions. Expressed as pH, recoveries ranged from 0.4
to 1.7 units.

Alkalinity and Base Cations

Alkalinity data were available only for Swan Lake and Lake 223.
In both cases, the alkalinity increase was even greater than the
decrease in for hydrogen ion. In Swan Lake, alkalinity was calculated
as the difference between base cations and strong acid anions. It was
necessary to assume that chloride and nitrate did not change during
recovery, because data for these two ions were not available. The
alkalinity increase, based on differences between base cations and
sulfate, was 188 $\mu eq \cdot L^{-1}$, twice the change observed in H^+ (Fig.
2). This appears to have happened because of slower decreases in base
cations than in strong acid anions.

A similar result was obtained in Lake 223, where a Gran
alkalinity increase of 14 $\mu eq \cdot L^{-1}$ accompanied a decrease in H^+ of
only 5 $\mu eq \cdot L^{-1}$. Base cations decreased by 13 $\mu eq \cdot L^{-1}$, while
sulfate decreased by 19 $\mu eq \cdot L^{-1}$. As in Swan Lake, the decrease in

sulfate relative to base cations appeared to be the mechanism of alkalinity generation, although these results must be regarded as tentative at present.

Trace Metals

Aluminum data were available only for Clearwater and Swan Lakes. Decreases were 56 and 75%, respectively (Table 3). Copper, nickel and zinc also decreased following reduction in emissions, but decreases in the first two elements are probably due largely to decreased deposition of the metals, rather than in-lake processes (P. Dillon, pers. comm.). Manganese decreased in Baby, Alice and Lake 223 (Table 3).

Biological Recovery

So far, complete biological information is not available for any of the lakes. I shall therefore present only a brief summary of results compiled so far, for Baby and Alice Lakes and Lake 223.

Phytoplankton diversity increased slightly in all three lakes during recovery. In Lake 223, diversity at pH 5.6 during recovery was similar to that observed when the pH of the lake was being decreased and many of the same algal species were present (D. Findlay, pers. comm.). In Baby and Alice Lakes, a considerable increase in phytoplankton biomass has also occurred (M. Havas, pers. comm.). There has also been an apparent increase in zooplankton in Alice Lake (chiefly Polyarthra vulgaris, Pleosoma trunacatum and Bosmina longirostris). Of course, the pH of this lake was never less than 6.0 during the period of the study. A more comprehensive analysis of the biota is under way (M. Havas, pers. comm.)

In Lake 223, Holopedium gibberum and Daphnia catawba decreased in number. This may be an indirect effect of the pH increase, caused by the return of planktivorous fishes (D. Malley, pers. comm.).

The increase in pH from 5.0 to 5.4 caused an immediate recovery of spawning by the white sucker (Catostomus commersoni) and pearl dace (Semotilus margarita) (K.H. Mills, pers. comm.). Both species had reproduced at this pH during the acidification phase of the study. In addition, a large number of brook sticklebacks (Culaea inconstans) invaded the lake. This species was rare prior to and during acidification, and it may have moved into one of the niches vacated by destroyed species. All three species were abundant in the lake and easily caught (K. Mills, pers. comm.).

Lake trout (Salvelinus namaycush) rapidly recovered in condition from their poor state at pH 5.0. They did not, however, have succesful recruitment in fall of 1984. So far it isn't known whether recruitment in 1985 was successful, because this depends on capturing young which hatched over winter in 1985-1986 (K.H. Mills, pers. comm.). So far, crayfish (Orconectes virilis), fathead minnows (Pimephales promelas), sculpins (Cottus cognatus) and opossum shrimp (Mysis relicta) have not returned to the lake, and the pH is still not high enough to support them. The lake will be monitored closely for their return during the recovery part of the experiment, and restocking will be done if it proves necessary.

How Predictable is the Recovery of an Acidified Lake?

The flushing time (τ_w) of lakes is known to be an important driving force in predicting the steady-state concentration of phosphorus in lakes, and is incorporated in many eutrophication models (Vollenweider 1969, 1975, 1976; Dillon 1975; Schindler et al. 1978). It is reasonable to expect that it would play an even more important role in determining the concentration of sulfur, which is much more conservative. Recently, Kelly et al. (under review) have shown water renewal to be an important predictor of the retention of both sulfate and nitrate in lakes, indicating that this assumption is correct.

However, several features of the sulfur cycle render such calculations difficult. In heavily polluted areas, standard methods for calculating dry deposition appear to greatly underestimate the input of sulfur from SO_2 (Jeffries et al. 1984). Also, studies of recovery in roofed acidified catchments show a considerable delay in sulfate losses (Wright and Gjessing 1986). On the other hand, sulfate reduction and sedimentation become an increasingly important feature as water renewal time increases. These errors should counterbalance each other to some degree, and there is generally good correspondence between sulfate losses and estimated sulfate inputs (Thompson 1982).

Water renewal times and long data records were available for Clearwater, Lohi, Middle and Hannah Lakes (Table 1, Dillon et al. 1986). It was assumed that the concentration of sulfate in all four lakes would be at steady state in 1976-1977, because deposition had been relatively constant for several years, and the water renewal times of all four lakes are quite short. We therefore assumed that the ratio of the steady-state concentration of sulfate in lakes under the current deposition regime would be the same ratio to current deposition as the concentration/deposition ratio in 1976-77.

The new steady-state concentration was assumed to be approached as a function of the water renewal rate:

$$(1) \qquad C_t = C_e + (C_o - C_e)e^{-\frac{1}{\tau_w}t}$$

where C_t is the concentration of sulfate at time t, C_o is the steady-state concentration prior to reduction, C_e is the new expected steady-state, and both τ_w and t are in years. C_e was calculated from the concentration/deposition ratio:

$$(2) \qquad C_e = \frac{D_{SO_4\ 84}\ C_o}{D_{SO_4\ 73-76}}$$

where D is the average annual deposition of sulfate in the years indicated, from Dillon et al. 1986, and values of τ_w were taken from Scheider (1984).

This simple model fit Clearwater Lake data almost exactly (Fig. 3a). In contrast, it overestimated the rate of recovery in the other three lakes, although after several years the lakes approached the predicted steady-state concentrations nevertheless (Fig. 3b). It may be that delays in "washout" of sulfate from the terrestrial watersheds caused sulfate removal to be slower than predicted in the first few years after deposition was reduced, or that the chemical manipulation of the lakes somehow affected the sulfate concentration. More information is needed about delays in terrestrial processes before more complex models are developed, but it appears that lake recoveries nevertheless occur more or less at rates predicted by decreases in deposition and water renewal times.

In conclusion, in eight lakes where inputs of sulfate were decreased, there were rapid decreases in sulfate. In the lakes where changes in H^+ could be attributed to acid inputs, large decreases in hydrogen ion concentration were observed. Similar observations have been made on other lakes near Sudbury (see discussion in Hutchinson and Havas 1986). Toxic trace metals also decreased dramatically. These observations lead us to be optimistic that when deposition is reduced sufficiently, lakes will rapidly return to a chemical state suitable for their natural biota.

Unassisted biological recovery seems less certain. Phytoplankton and many zooplankton appear to respond quickly. Many of the fish species which have not been completely eliminated appear to be capable of resuming reproduction quickly once lake conditions are suitable. Unfortunately, data to date do not allow us to include the lake trout, highly valued as a sport fish, among these. Species with low powers of dispersal which have been eliminated from the lakes, such as Mysis and Orconectes may not return unless restocked.

It is also possible that species eliminated by acidification may not be able to recover their vacated niches from species which invaded while they were absent, even when the chemistry of lakes has recovered. For example, it remains to be seen whether the fathead minnow, Pimephales promelas will return to displace the brook stickleback. Thus, while the chemical recovery of lakes seems to occur at a predictable rate, the biological recovery is not predictable at present. Until the extent of biological recovery is known, we have no way of assessing what the long-term costs of damage from acid rain to ecosystems has been.

4. ACKNOLWEDGEMENTS

Ken Beaty, Dave Findlay, Dana Cruikshank, Ian Davies, Garry Linsey, Diane Malley and Ken Mills contributed unpublished information for this summary. Michael Turner helped to assemble data and tables, and performed calculations for the chemical budgets. Peter Dillon and Magda Havas supplied pre-publication copies of their papers. D. Laroque assembled the figures and typed the manuscript.

REFERENCES

(1) BAYLEY, S.E. et al. (1986). Experimental acidification of a Sphagnum-dominated peatland: first year results. Can. J. Fish. Aquat. Sci. (In press)

(2) BEAMISH, R.J. and Harvey, H.H. (1972). Acidification of the LaCloche Mountain lakes, Ontario, and resulting fish mortalities. J. Fish. Res. Board Can. 29: 1131-1143.

(3) CONROY, N. et al. (1974). Acid Shield lakes in the Sudbury, Ontario region. Proc. 9th Can. Symp. Wat. Pollut. Res. Can., No. 9, p. 45-61.

(4) COOK, R.B. et al. (1986). Mechanisms of hydrogen ion neutralization in an experimentally acidified lake. Limnol. Oceanogr. 31: 134-148.

(5) DILLON, P.J. (1975). The phosphorus budget of Cameron Lake, Ontario: the importance of flushing rate to the degree of eutrophy of lakes. Limnol. Oceanogr. 20: 28-39.

(6) DILLON, P.J., REID, R.A. and GIRARD, R. (1986). Changes in the chemistry of lakes following reductions of SO_2 emissions. Water Air Soil Pollut. (In press)

(7) GORHAM, E. and GORDON, A.G. (1960). The influence of smelter
     fumes upon the chemical composition of lake waters near Sudbury,
     Ontario and upon surrounding vegetation. Can. J. Bot. 38:
     447-487.
(8) HUTCHINSON, T.C. and HAVAS, M. (1986). Recovery of previously
     acidified lakes near Coniston, Canada following reductions in
     atmospheric sulphur and metal emissions. Water Air Soil Pollut.
     (In press).
(9) HUTCHINSON, T.C. and WHITBY, L.M. (1977). The effects of acid
     rainfall and heavy metal particulates on a boreal forest
     ecosystem near the Sudbury smelting region of Canada. Water Air
     Soil Pollut. 7: 421-438.
(10) JEFFRIES, D.S. et al. (1984). Atmospheric deposition of
     pollutants in the Sudbury area, p. 117-154. In J. Nriagu (ed.)
     Environmental Impacts of Smelters. John Wiley and Sons, Inc.,
     New York.
(11) KELLY, C.A. et al. Prediction of biological acid neutralization
     in lakes. Biogeochemistry. (Under review)
(12) SCHEIDER, W.A. (1984). Lake water budgets in areas affected by
     smelting practices near Sudbury, Ontario, p. 155-193. In J.
     Nriagu (ed.) Environmental Impacts of Smelters. John Wiley and
     Sons, Inc., New York.
(13) SCHINDLER, D.W. (1985). The coupling of elemental cycles of
     organisms: evidence from whole lake chemical perturbations, p.
     225-250. In W. Stumm (ed.) Chemical Processes in Lakes. John
     Wiley and Sons, Inc., New York.
(14) SCHINDLER, D.W. (1986). The significance of in-lake production
     of alkalinity. Water Air Soil Pollut. (Under review)
(15) SCHINDLER, D.W. et al. (1985). Long-term ecosystem stress: the
     effects of years of acidification on a small lake. Science
     (Wash., DC) 228: 1395-1401.
(16) SCHINDLER, D.W. et al. (1986). Natural sources of acid
     neutralizaing capacity in low alkalinity lakes of the Precambrian
     Shield. Science 232: 844-847.
(17) SCHINDLER, D.W. and TURNER, M.A. (1982). Biological, chemical
     and physical responses of lakes to experimental acidification.
     Water Air Soil Pollut. 18: 259-271.
(18) SCHINDLER, D.W., FEE, E.J. and RUSZCZYNSKI, T. (1978).
     Phosphorus input and its consequences for phytoplankton standing
     crop and production in the Experimental Lakes Area and in similar
     lakes. J. Fish. Res. Board Can. 35: 190-196.
(19) THOMPSON, M.E. (1982). Denudation rate as a quantitative index
     of sensitivity of eastern Canada rivers to acidic atmospheric
     precipitation. Water Air Soil Pollut. 18: 215-226.
(20) VOLLENWEIDER, R.A. (1969). Moglichkeiten and Grenzen
     elementarer Modelle des Stoffbilanz von Seen. Arch. Hydrobiol.
     66: 1-36.
(21) VOLLENWEIDER, R.A. (1975). Input-output models with special
     reference to the phosphorus loading concept in limnology.
     Schweiz. Zerts. Hydrol. 37: 53-84.
(22) VOLLENWEIDER, R.A. (1976). Advances in defining critical
     loading levels for phosphorus in lake eutrophication. Mem. Ist.
     Ital. Idrobiol. Dott Marco de Marchi 33: 53-83.

Lake	Area, ha		Depth, m		τ_w, yr	Distance from Sudbury, km	Basin characteristics and surficial geology
	Terrestrial drainage	lake	max	$\bar{x}$			
Baby	59.3	11.7	not available		not available	1.0	granites, gneisses – no soil or vegetation
Alice	316	26.7	not available		not available	0.6	deep tills
Clearwater	340	76.5	21.5	8.3	3-4	13	85 to 90% exposed gneiss, quartzite, gabbro; remainder is peat covered, plus one small pond
Middle	250	28.2	15.0	6.2	1.2-1.6	5	75-80% exposed quartzite and gabbro; remainder is thin tills and clay
Hannah	76	27.3	8.5	4.0	2.1-2.9	4	
Lohi	460	40.5	19.5	6.2	0.9-1.2	11	56% exposed quartzite and gabbro; remainder is thin till
Swan	n.d.	5.8	8.8	2.8	n.d.	13	n.d.
223	135.2	27.3	14.6	7.2	2-20	>1000	exposed gneisses, some peat-covered; pockets of thin till

Table 1 : Morphometric and geographical characteristics of the study lakes

Table 2 : Concentrations of sulfate, pH, conductivity and alkalinity available for the study lakes, before and after the reduction in emissions

Lake	SO$_4$, μeq·L^{-1}			pH			Conductivity, μmhos·cm^{-1}			Alkalinity	
	before	after	% decrease	before	after	% decrease[1]	before	after	% decrease	before	after
Clearwater[4]	545	370	32	4.23	4.61	59	89	79	11	-59	-31
Middle[4,6]	850	560	34								
Hannah[4,6]	1150	640	44								
Lohi[4]	525	390	25								
Swan[4]	580	220	62	3.96	4.80	85	98	48	51	-66	122[3]
Baby[5]	1499	624	58	4.1	5.8	98	266	92	65		
Alice[5]	5538	1915	65	6.1	6.8	80	490	250	49		
223[2]	228	202	13	5.18	5.67	67	36.9			- 4	10

1 Decrease in (H$^+$)
2 Data are for all overturn in 1983 and 1985
3 Alkalinity is calculated by ion balance, ignoring NO$_3^-$ and Cl$^-$
4 From Dillon et al. 1986
5 From Hutchinson and Havas 1986
6 Treated with CaCO$_3$ and/or Ca (OH)$_2$

Table 3 : Trace metal concentrations in the study lakes, before and after decreases in deposition of SO$_2$ and trace metals. Data in μg. L^{-1}

Lake	Al			Cu			Ni			Zn			Mn		
	before	after	% decrease	before	after	% decrease	before	after	% decrease	before	after	% decrease	before	after	% decrease
Clearwater	430	190	56	100	52	48	290	220	24	50	24	52			
Swan	290	63	78	64	12	81	300	73	76	36	13	63			
Baby	n.d.	130	-	780	60	92	3200	410	87	180	60	67			
Alice	n.d.	760	-	250	9	96	6900	1400	80		no trend				
223													155	93	40

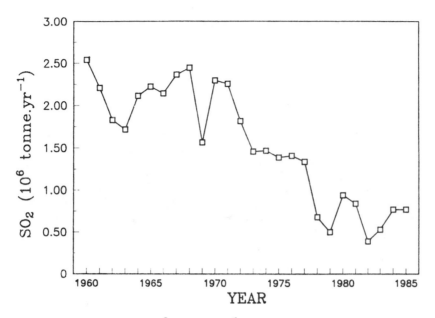

Figure 1 : SO$_2$ emissions (10^6 tonne.yr^{-1}) in the Sudbury basin (1960–1985)
From Dillon et al. (1986)

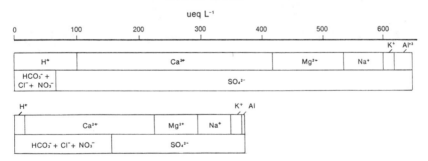

Figure 2 : A bar graph illustrating the ionic changes responsible for the
increase in alkalinity of Swan Lake, near Sudbury. The upper bar
is for 1977, the lower for 1982. Data from Table 1 in Dillon et
al. (1986)

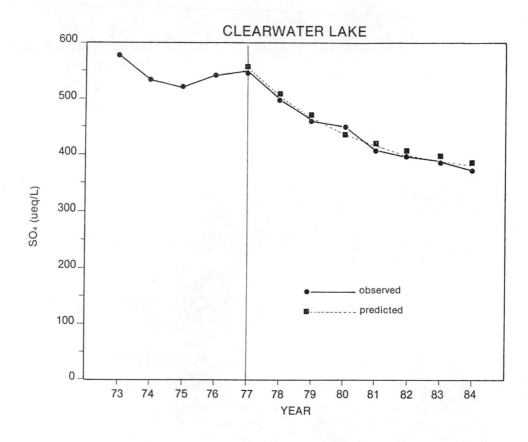

Figure 3a : Clearwater Lake. Actual (solid lines) and theoretically predicted (dotted lines, equation 1) reductions in sulfate in lakes near Sudbury. Predictions were made from 1976. Figures modified from Dillon et al. (1986). Water renewal times used in equation 1 from Scheider (1984)

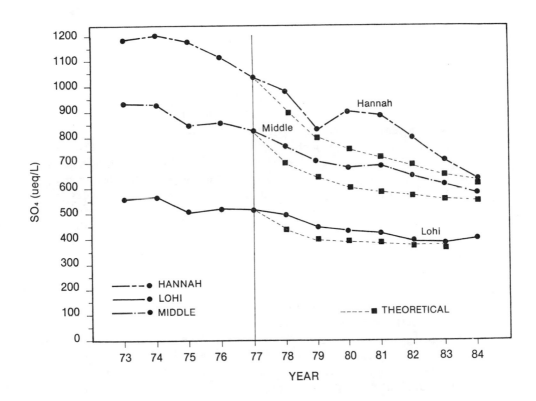

Figure 3b : Middle, Hannah and Lohi Lakes. Actual (solid lines) and theoretically predicted (dotted lines, equation 1) reductions in sulfate in lakes near Sudbury. Predictions were made from 1976 onward. Figures modified from Dillon et al. (1986). Water renewal times used in equation 1 from Scheider (1984).

RAIN PROJECT: RESULTS AFTER 2 YEARS OF TREATMENT

R.F. WRIGHT
Norwegian Institute for Water Research,
P.O. Box 333, 0314 Oslo 3, Norway

Summary

Project Rain (Reversing Acidification In Norway) is a 5-year
international research project aimed at investigating the effect on
water and soil chemistry of changing acid deposition to whole
catchments.   The project comprises 2 parallel large-scale
experimental manipulations -- artificial acidification at Sogndal and
exclusion of acid rain at Risdalsheia. Treatment at Sogndal commenced
April 1984 with the acidification of the snowpack by addition of
sulfuric acid (SOG2) and a 1:1 mixture of sulfuric and nitric acids
(SOG4). Acid addition caused rapid and dramatic increases in runoff
acidity and aluminum concentrations. Most of the added acid was
retained in the catchments. At Risdalsheia treatment began in June
1984 at KIM catchment (treatment by deacidified rain) and EGIL
catchment (control with ambient acid rain). After 1 1/2 years of
treatment about 1100 mm of clean precipitation had been applied at
KIM. Nitrate is reduced by 75% and sulfate by 35% in runoff from KIM
relative to EGIL. Reduction in these anions were balanced by lower
levels of most cations. Aluminum concentrations in runoff at Sogndal
indicated a lower solubility of labile Al in 1985 relative to 1984.
Acid addition has perhaps depleted a reservoir of readily-soluble
aluminum in the streambed at these catchments.

1. INTRODUCTION

Vigorous efforts to obtain reductions in the emissions of acidifying
compounds SO_2 and NOx to the atmosphere are in part based on the premise
that such reductions will restore acidified waters. The magnitude and
rate of response of natural ecosystems to changes in acid loading is,
however, not well known, largely because such effects have been difficult
to document in the absence of large-scale reductions. We are now
conducting manipulations of natural headwater catchments in Norway.
Project RAIN (Reversing Acidification In Norway), a 5-year international
research project, comprises two parallel experiments in which the response
of soil and runoff chemistry to changes in loading of strong acids from
the atmosphere are studied (Wright 1985, Wright and Gjessing 1986, Wright
et al. in press). The RAIN project builds on short-term pilot-scale

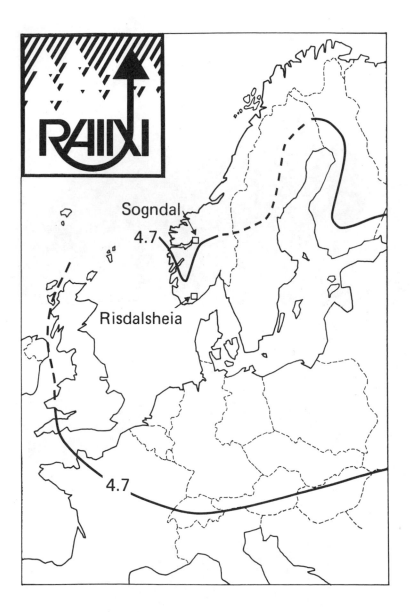

Figure 1. Location of the experimental catchments in project RAIN. Areas within the pH 4.7 isoline receive precipitation with a yearly weighted-average pH below 4.7.

Figure 2. Sogndal. Application of acid to catchment SOG2 in September 1984. The weir at outlet to the small pond is at lower right.

Figure 3. Risdalsheia. Mounting of roof panels on EGIL catchment in June 1984.

experiments conducted in Norway by Seip et al.(1979) and Christophersen et al. (1982).   The project provides information on reversibility of acidification, rate of response and target loadings.

At Sogndal, a "clean" area in western Norway (Figure 1), we are acidifying two pristine catchments by addition of sulfuric (SOG2) and a 1:1 mixture of sulfuric and nitric acids (SOG4), respectively (Figure 2, Table 1). At Risdalheia, an acidified area in southernmost Norway (Figure 1), we have excluded acid precipitation from a small catchment (KIM) by means of a roof and are watering with clean precipitation beneath the roof (Figure 3, Table 1).

The RAIN project design, organization, site descriptions and results are described in the annual reports for 1984 (Wright 1985) and 1985 (Wright and Gjessing 1986). The first year's results were presented at the Muskoka '85 conference (Wright et al. in press).

Table 1. RAIN project. Overview of the experimental catchments and treatments.

	Sogndal. Acid addition experiments.	
Catchment	Treatment	Area
SOG1	control	96300 m^2
SOG2	H_2SO_4	7220 m^2
SOG3	control	43200 m^2
SOG4	$H_2SO_4 + HNO_3$	1940 m^2

	Risdalsheia. Acid exclusion experiments.	
Catchment	Treatment	Area
KIM	roof, clean rain	860 m^2
EGIL	roof, acid rain	400 m^2
ROLF	no roof, acid rain	220 m^2

2. SITE DESCRIPTION

Sogndal.   For the acidification experiments we selected 4 pristine, headwater catchments at Sogndal. The site is located above treeline at about 900 m above sealevel. Vegetation is alpine and includes dwarf birch, heather, grasses, mosses and lichens. Soils are thin (average depth about

30 cm) and poorly-developed (US classification: Lithic Haplumbrept, sandy, siliceous, frigid). Soil pH (H_2O) is 4.5-5.5. Bedrock is siliceous gneiss (Lotse and Otabbong 1985). Precipitation at Sogndal is relatively clean with a volume-weighted average pH of 4.9 and excess sulfate concentration of 14 µeq/l. Ambient loading of excess sulfate is about 9 kg/ha/yr. Runoff chemistry from the 4 catchments prior to treatment indicates extremely sensitive systems with pH 5.5-6.0, Ca concentrations 10-30 µeq/l, SO_4 15-25 µeq/l, and alkalinity 5-15 µeq/l.

Risdalsheia. For the de-acidification experiments we selected 3 natural headwater catchments at Risdalsheia, located about 5 km from the long-term calibrated catchment at Birkenes. The site is 300 m above sealevel. Vegetation at Risdalsheia is similar to that at Sogndal except that here pine and birch trees are also present. The soils are similar -- thin (average depth 15 cm), patchy and poorly-developed. A major difference is that here soil pH (H_2O) is about one unit lower (pH 3.9-4.5). Bedrock is biotite granite. Risdalsheia lies in the zone of maximum deposition of acid components in Norway. Data from the nearby Birkenes station show a long-term volume-weighted average pH of 4.2 in wet precipitation (Christophersen and Wright 1981). Deposition of sulfate is 70 kg/ha/yr (50 wet plus 20 dry). Runoff from the 3 catchments prior to treatment was highly acidic (pH 3.9-4.2) with high levels of sulfate and inorganic labile aluminum.

3. TREATMENTS

Treatment at Sogndal commenced in April 1984 with the addition of acid to the snowpack. About 0.02 mm of pH 1.9 acid was spread evenly over catchments SOG2 and SOG4. Prior to treatment bulk snow had a pH of about 4.9; after acid addition pH was 4.2. A commercial sprinkling system was used to apply acid in four separate episodes during August-October 1984. The acid was added to lakewater from SOG1; 11 mm of pH 3.2 water was added. Before and after each treatment 2 mm of pH 5.6 lakewater was added to wet up and wash the vegetation, respectively. Watering intensity is 2 mm/hr. No direct damage to the vegetation was observed. Details are given by Wright (1985). Treatment continued with acid addition to the snowpack in April 1985 and watering in June - October 1985. A total of 151 keq H^+/km^2 was added during the period April 1984 - October 1985.

At Risdalsheia the experiment includes 3 headwater catchments (Table 1). Two are covered by roofs (KIM treatment; EGIL control) and one (ROLF) serves as untreated reference. The roofs extend 2-3 meters beyond the catchment boundaries. Treatment at KIM and EGIL catchments entails collection of incoming precipitation by means of gutter and cistern systems. At KIM the water is sent through ion-exchange columns. Seawater is added to give the natural levels of seawater salts. The resulting precipitation has a chemical composition equivalent to that at Birkenes except that H^+, excess SO_4, NO_3, and NH_4 are removed. The water is pumped back out to a sprinkler system mounted beneath the roof. Watering proceeds at 2 mm/hr. The system at EGIL catchment is similar except that here the water is not treated but merely recycled back beneath the roof. The entire system is automatic and regulated by the water level in the cisterns. The system capacity is about 50 mm/day. During the winter we shut down the watering systems, lower side panels to prevent snow from blowing into the catchments, and make artificial snow using commercial snow-making equipment. Details are given by Wright (1985).

4. RESULTS

At Sogndal both catchments responded immediately and dramatically to the acid treatments (Figure 2). During snowmelt in 1984 runoff at SOG2 was acidified to pH 4.1 and contained high levels of sulfate and labile aluminum. Due to technical difficulties no samples from SOG4 were collected during snowmelt in 1984. During the 4 episodes of acid addition in August-October 1984 pH levels fell to 4.7 at SOG2 and 4.3 at SOG4. The dominant anion was sulfate at SOG2 and approximately equivalent concentrations of sulfate and nitrate at SOG4 (Figure 4). At both catchments acidic runoff contained high concentrations of labile Al.

Whereas SOG4 responded and returned to background levels within hours, the response at SOG2 is considerably damped. Runoff chemistry exhibits a step-like pattern with increasingly higher acidity, sulfate and labile Al concentrations following each of the four treatments (Figure 4). The small pond at the bottom of SOG2 provides a hydrologic buffer that apparently acts to smooth out sudden changes in runoff chemistry such as observed at SOG4.

Of the 70 keq SO_4/km^2 added to SOG2 in 1984 approximately 20 % left in runoff (Table 2). Only 10 % (7 keq/km^2) of the H$^+$ left in runoff, however, with the difference accounted for by increased labile aluminium and base cations and decreased bicarbonate alkalinity. At SOG4 about 10 % of the added SO_4 and 5 % of the added NO_3 left in runoff. In 1985 a somewhat larger fraction of the added sulfate (83 keq/km^2) left SOG2 in runnoff. At SOG4, however, less SO_4 and less NO_3 left in 1985 relative to 1984 (Table 2). Catchment SOG2 did not recover fully to background levels of sulfate during the winter of 1984-85 (Figure 4). Sulfate levels during 1985 were consistently higher than those at SOG3, the untreated control catchment. Several of the acid additions in 1985 coincided with major precipitation events, and thus the peaks in sulfate, nitrate, labile aluminum and H$^+$ were not as pronounced in 1985 as compared with the previous year.

At Risdalsheia nitrate concentrations in runoff at KIM catchment declined promptly in response to treatment from about 10-20 µeq/l to below detection limit (Figure 5). Sulfate concentrations in runoff began to deviate relative to levels at EGIL and ROLF catchments in mid-October, at which time about 350 mm of precipitation had been ion-exchanged. At this

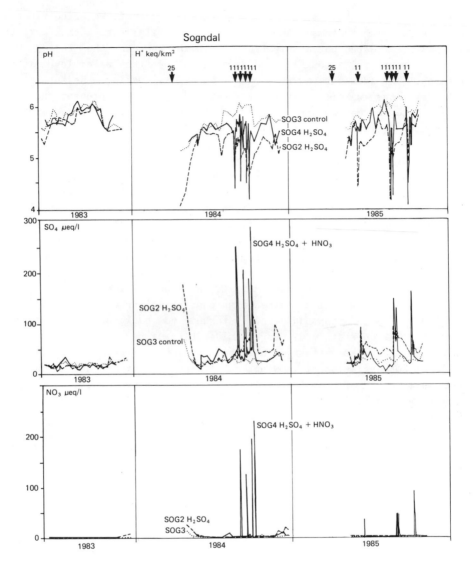

Figure 4. pH and concentrations of sulfate and nitrate in runoff from catchments SOG2, SOG3 and SOG4 at Sogndal over the period June 1983 – June 1985. Treatments began in April 1984 with the acid addition to the snowpack. Acid additions are indicated by arrows.

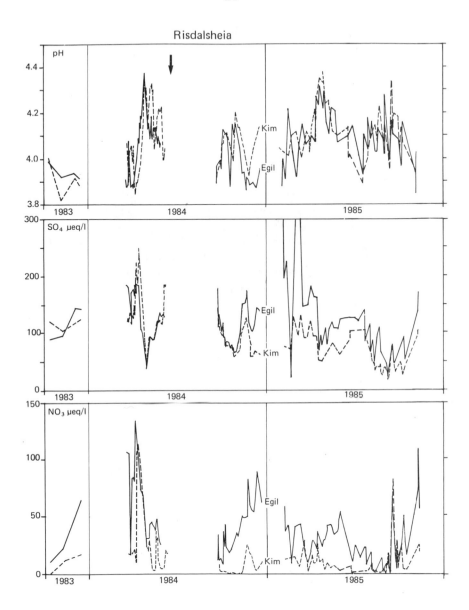

Figure 5. pH and concentrations of sulfate and nitrate in runoff from KIM and EGIL catchments at Risdalsheia over the period October 1983 - June 1985. Treatment began in mid-June 1984 (arrow).

time pH levels began to go up and levels of labile Al decreased. By mid-December 1984 at the onset of winter sulfate concentrations at KIM were about 80 µeq/l as compared to 100-120 µeq/l at EGIL and ROLF, and pH levels were about 0.2 units higher.

The trends in runoff water quality observed in late 1984 at KIM catchment continued in 1985. Nitrate levels remained significantly lower in runoff from KIM than that from both EGIL and ROLF (Figure 5). Sulfate levels from KIM continued at about 60-80 µeq/l whereas runoff from EGIL catchment contained significantly higher levels (90-140 µeq/l).

Sulfate levels in runoff from EGIL continued to exhibit seasonal variations related to deposition and accumulation in the catchment (Figure 5). At KIM, however, the sulfate maxima no longer occur. Concentrations

Table 2. Input-output budgets for water and major ions at the Sogndal catchments for the periods 831115-841113 (1984) and 841114-851101 (1985). Units: H_2O in mm; ions in meq/m^2/yr.; TOC in mg C/m^2/yr; SiO_2 in mg SiO_2/m^2/yr.

- 1984 -

	SOG1 control		SOG2 H_2SO_4		SOG3 control		SOG4 H_2SO_4 + HNO_3	
	In	Out	In	Out	In	Out	In	Out
H_2O	1126	1026	1186	1086	1126	1026	1186	1086
H^+	19	2	89	9	19	3	89	4
Na	70	90	73	64	70	89	73	55
K	6	11	6	4	6	2	6	2
Ca	9	21	10	24	9	22	10	28
Mg	16	15	16	14	16	18	16	12
Al	–	1	–	3	–	1	–	2
NH_4	9	6	9	3	9	1	9	1
NO_3	9	2	9	4	9	2	44	4
Cl	79	91	82	65	79	85	82	62
SO_4	30	29	100	43	30	25	65	29
HCO_3	0	15	0	5	0	2	0	7
Σ^+	129	146	203	121	129	137	203	104
Σ^-	119	138	191	113	119	114	192	102
TOC	–	4.4	–	2.7	–	0.8	–	2.9
SiO_2	–	–	–	–	–	–	–	–

Table 2. (cont.)

- 1985 -

	SOG1 control		SOG2 H_2SO_4		SOG3 control		SOG4 H_2SO_4 + HNO_3	
	In	Out	In	Out	In	Out	In	Out
H_2O	928	828	991	891	928	828	991	891
H^+	19	2	102	4	19	2	102	3
Na	41	29	43	28	41	27	43	27
K	6	2	6	2	6	1	7	2
Ca	5	13	5	20	5	13	6	18
Mg	5	7	5	9	5	7	5	8
Al	–	0	–	2	–	0	–	1
NH_4	8	4	8	0	8	1	8	1
NO_3	10	1	10	1	10	1	51	1
Cl	33	24	35	23	33	20	31	22
SO_4	21	18	104	35	21	18	63	20
HCO_3	0	11	0	2	0	5	0	5
Σ^+	84	58	169	65	84	51	170	61
Σ^-	64	54	149	61	64	44	144	49
TOC	–	1.3	–	1.4	–	0.8	–	2.5
SiO_2	–	0.8	–	1.2	–	0.8	–	1.6

were relatively constant during 1985. Desorption of sulfate from the soils is a process by which sulfate levels in runoff could be kept relatively constant at concentrations higher than inputs.

By 13 November 1985 after 1 1/2 years of treatment a total of 1100 mm of rain and snow have been ion-exchanged and applied beneath the roof at KIM catchment. At EGIL catchment 1350 mm have been applied. During this same period natural precipitation amounted to about 2270 mm. Average-annual precipitation at Birkenes is 1350 mm which is also our best estimate for Risdalsheia.

Thus although treatment has proceeded for 1 1/2 calendar years, the equivalent of only about 1 year of precipitation has been applied to KIM and EGIL. The difference is due to periods with technical problems during which all incoming precipitation is not collected, periods of high precipitation intensity during which the designed capacity of the sprinkler system is exceeded, and the much larger natural snowfall winter 1985 (550 mm) as compared with the 115 mm of artificial snow produced.

Input-output budgets for KIM and EGIL catchments illustrate the effects of reduced deposition at KIM. The major features of input-outputs budgets for EGIL catchment are (1) approximate balance of H+ and base cations, (2) net output of Al, (3) large net uptake of NH_4 and NO_3, and (4) approximate balance of SO_4 (Table 3). Cl was assumed to balance. With the exception of the input-output balance for H+ and base cations these results are similar to those generally found at small catchments receiving acid precipitation in Norway (Wright and Johannessen 1980). That base cation outputs equal inputs is due to the extremely thin soils and the large fraction of the catchment at EGIL and KIM with exposed bedrock. Weathering rates are thus low.

Table 3. Input-output budgets for water and major ions at EGIL and KIM catchments for the $1^1/_2$ year treatment period 13 June 1984 - 14 November 1985. Units: meq/m^2. See text for details.

	EGIL						KIM			
		—Input—			Total	Out	—Input—			Out
	Wet	—Dry—					Wet	Dry	Total	
		marine part.	Acid part.	gases						
H_2O (mm)	1346					1092	1102			835
H^+	64	0	2	31	97	102	10	33	43	72
Na	65	32	0	0	97	91	61	32	93	77
K	6	1	0	0	7	8	1	1	2	6
Ca	12	1	0	0	13	15	3	1	4	11
Mg	16	7	0	0	23	24	14	7	21	14
Al	0	0	0	0	0	18	0	0	0	11
NH_4	57	0	6	0	63	16	0	6	6	6
NO_3	50	0	0	18	68	32	1	18	19	6
Cl	70	38	0	0	108	108	72	38	110	104
SO_4	86	4	8	13	111	115	8	25	33	58
Org.anion	0	0	0	0	0	18	0	0	0	28
Σ^+	220	41	8	31	300	274	89	80	169	197
Σ^-	206	42	8	31	287	273	81	81	162	196

About 75% of the incoming NH_4 and 50% of the incoming NO_3 are retained in the catchment (Table 3). NH_4 retention results from both biological uptake as well as ion exchange. NO_3 is retained by biological activity, but ion exchange is relatively unimportant for this mobile anion. The balanced SO_4 budget is an indication that the estimated dry deposit is approximately correct.

At KIM catchment the 1 1/2 year treatment has caused changes in the flux of H+, base cations, Al, NH_4, NO_3, and SO_4 (Table 3). Whereas EGIL catchment showed approximately equal flux of H+ out as in, reduction of H+ inputs at KIM have resulted in a net flux of H+ out of the catchment. Of the base cations both Na and Mg now show net accumulation in KIM catchment. With the lower concentrations of mobile anions SO_4 and NO_3 at KIM, concentrations of cations must also decrease. The result may be a build up of the pool of exchangeable base cations on the soil. The net flux of Al is also lower at KIM than at EGIL, perhaps because of the lower acid deposition and slightly higher pH levels in runoff at KIM due to treatment.

Treatment at KIM caused major decreases in NH_4 and NO_3 input, and output responded as well (Table 3). NH_4 deposition at KIM is only 10% of that at EGIL, and NH_4 outputs have declined such that output equals input. NO_3 inputs have also decreased, but not to such a great extent due to continued dry deposit of NO_2 gas. NO_3 outputs at KIM are now only 20% of those at EGIL. Whereas NH_4 and NO_3 were the third most important cation and anion, respectively, in both deposition and runoff at EGIL, at KIM they now are minor consituents.

The sulfate budget at KIM shows major differences from that of EGIL (Table 3). Whereas at EGIL SO_4 out equals SO_4 in, at KIM there is almost twice as much SO_4 leaving the catchment as enters in dry and wet deposition during the first 1 1/2 years of treatment.

The SO_4 budget indicates a net loss of about 25 meq/m^2 from KIM. The measurements of Lotse and Otabong (1985) of water soluble SO_4, adsorbed SO_4 and SO_4 in soil solution yield an estimate of about 40 meq/m^2 SO_4 in the catchment. Thus about 2/3 of the readily-available SO_4 has been lost from KIM as of November 1985.

5. DISCUSSION

Acid additions at Sogndal in 1984 caused increases in labile Al from natural levels of <50 µg Al/l to as much as 1000 µg Al/l (Figure 6). The data from 1984 indicated that an $Al(OH)_3$ solid phase with pK of about 9.1 can account for labile Al concentrations at Sogndal (Figure 6) (Wright 1985). This solubility corresponds to that of synthetic gibbsite at $10^0 C$ or natural gibbsite at $20^0 C$ (Driscoll et al. 1984, Seip et al. 1984). Similar Al-pH relationships are found elsewhere in Norway (Wright and Skogheim 1983, Seip et al. 1984) and in North America (Driscoll 1980, Driscoll et al. 1984, Johnson et al. 1981).

The data from the second year of treatment (1985) indicate a significantly lower aluminum solubility with pK of about 7.7 (Figure 6). We postulate that the acid addition during 1984 at SOG2 and SOG4 resulted in the depletion of a reservoir of readily-soluble $Al(OH)_3$ with pK of 9.1. Al precipitated in the stream channels may provide such a reservoir. We

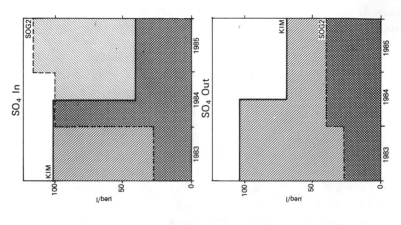

Figure 7 : Volume weighted-average concentrations of sulfate in deposition and runoff at SOG2 (Sogndal) and KIM (Risdalsheia) before and after treatment. Concentrations in deposition are obtained by dividing input flux (wet and dry) by runoff volume thus adjusting for evapotranspiration

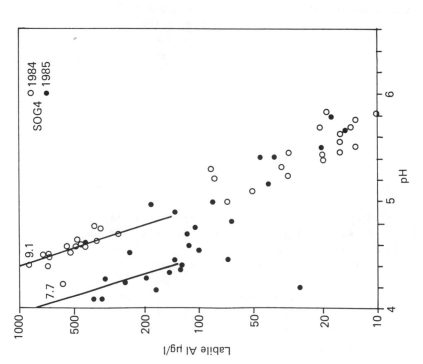

Figure 6 : Concentrations of labile aluminium and pH in runoff samples from Sogndal in 1984 (open circles) and 1985 (solid circles). Also shown are lines of equilibrium with $Al(OH)_3$ of pK (expressed as $3pH-pAl^{+3}$) of 7.7 and 9.1

suggest that this Al, perhaps amorphous, occurs as a result of the natural increase in water pH as soil solution enters the stream and CO_2 degasses (Reuss and Johnson 1986). Our additions of strong acid in 1984 would then redissolve this $Al(OH)_3$ phase. By 1985 the reservoir is depleted, and Al concentrations in streamwater now reflect the Al solubility in soil solution. Here the Al comes from ion-exchange as well as weathering of primary and secondary minerals.

Labile aluminum levels at the highly acidified sites at Risdalsheia indicate pK of about 6.8 (Wright 1985). It appears that Al solubility at the Sogndal catchments is moving towards the situation at Risdalsheia.

Sulfate concentrations in runoff have responded differently to acid additions at Sogndal relative to acid exclusion at Risdalsheia. At the untreated control catchments at Sogndal the sulfate flux in equals the sulfate flux out. Concentrations in streamwater are about 25 µeq/l. Acid addition to SOG2 increased the volume-weighted concentration in precipitation to about 100 µeq/l the first year and to about 120 µeq/l the second year. Runoff sulfate responded by increasing slightly to 40 µeq/l both years (Figure 7). Most of the sulfate is retained in the catchment, presumably by sulfate adsorption.

At Risdalsheia the control catchment EGIL also shows a balanced sulfate budget with inputs (wet and dry) equivalent to about 100 µeq/l in runoff and outputs about the same (Figure 7). The exclosure at KIM catchment resulted in reduction of inputs to the equivalent of about 40 µeq/l. Runoff response during the first 1 1/2 years has been a decline to about 70 µeq/l. The net loss of sulfate can be explained by the desorption of water-soluble sulfate in the soils, although other processes in the sulfur cycle such as mineralization and transformation from organic-S can also account for these data (Seip et al. 1979).

Response at Risdalsheia has thus been more rapid than that at Sogndal This is consistent with the responses expected if the process of sulfate adsorption dominates in these soils. The response to an increase in sulfate deposition will be retention followed by a rapid "breakthrough", whereas the response to a decrease in deposition will be first rapid and then followed by a long "tail". This hysteresis is described by Reuss and Johnson (1986).

The RAIN project provides insight into the response of whole catchments to changes in loading of acids from the atmosphere. Although many of the results presented here are preliminary, we are hopeful that the RAIN project will help us understand the complex relationship between soil and water acidification and provide information useful for the establishment of target loadings and goals for future levels of emissions of acidifying compounds.

6. ACKNOWLEDGEMENTS

A large number of individuals and institutes have cooperated in the

RAIN project. The project scientists include N. Christophersen, E. Lotse, E. Gjessing, H.M. Seip, A. Semb, and R. F. Wright. Technical staff includes S. Andersen, H. Efraimsen, R. Høgberget, A. Rogne, B. Sletaune, R. Storhaug, and K. Wedum. NILU, NIVA, and SI and the Department of Soil Sciences, SLU, provided technical support.

The RAIN project would not be possible without the generous cooperation of landowners at both sites. We thank N. Knagenhjelm, Sogn Televerk and Arendal Televerk for permission to use private roads. N. Dalaker, H. Haukås, and A. Risdal provided local assistance.

Financial support has come from the Norwegian Ministry of Environment, The Royal Norwegian Council for Scientific and Industrial Research, the Ontario Ministry of the Environment, Environment Canada, the Swedish National Environmental Protection Board, and the Surface Water Acidification Programme (SWAP) (The Royal Society, the Norwegian Academy of Science and Letters, and the Royal Swedish Academy of Sciences).

REFERENCES

CHRISTOPHERSEN, N., STUANES, A.O., and WRIGHT, R.F. 1982. Runoff chemistry at a mini-catchment watered with "unpolluted precipitation". Nordic Hydrol. 13: 115-128.

CHRISTOPHERSEN, N., and WRIGHT, R.F. 1981. Sulfate flux and a model for sulfate concentrations in streamwater at Birkenes, a small forested catchment in southernmost Norway. Water Resour. Res. 17: 377-389.

DRISCOLL, C.T. 1980. Chemical characterization of some dilute acidified lakes and streams in the Adirondack Region of New York State. Thesis, Cornell Univ., Ithaca, NY.

DRISCOLL, C.T., BAKER, J.P., BISGONI, J.J., and SCHOFIELD, C.L. 1984. Aluminum speciation and equilibria in dilute acidic surface waters of the Adirondack Region of New York State, p.55-76, In Bricker, O.P. (ed.), Geological Aspects of Acid Deposition, Butterworth, Boston.

JOHNSON, N.M., DRISCOLL, C.T., EATON, J.S., LIKENS, G.E., and McDOWELL, W.H. 1981. 'Acid rain', dissolved aluminum and chemical weathering at the Hubbard Brook Experimental Forest, New Hampshire. Geochim. Cosmochim. Acta 45: 1421-1437.

LOTSE, E., and OTABBONG, E. 1985. Physiochemical properties of soils at Risdalsheia and Sogndal: RAIN project. Acid Rain Res. Rept. 8/85 (Norwegian Inst. Water Research, Oslo, Norway), 48pp.

REUSS, J.O., and JOHNSON, D.W. 1986. Acid deposition, soils, and water: an analysis. (Springer-Verlag, New York).

SEIP, H.M., GJESSING, E.T., and KAMBEN, H. 1979. Importance of the composition of the precipitgation for the pH in runoff-experiments with artificial precipitation on partly soil-covered "minicatchments". Internal Report IR 47/79, (SNSF-project, 1432 Ås, Norway), 34 pp.

WRIGHT, R.F. 1985. RAIN project. Annual report for 1984. Acid Rain Res. Rept. 7/85 (Norwegian Inst. Water Research, Oslo, Norway), 39pp.

WRIGHT, R.F., and GJESSING, E.T. 1986. Rain project. Annual report for 1985. Acid Rain Res. Rept. 9/85 (Norwegian Inst. Water Research, Oslo, Norway), 33pp.

WRIGHT, R.F., and JOHANNESSEN, M. 1980. Input-output budgets of major ions at gauged catchments in Norway, p. 250-251 In A. Tollan and D. Drabløs (Eds.) Ecological Impact of Acid Precipitation (SNSF-project, 1432 Ås, Norway), 383 pp.

WRIGHT, R.F., and SKOGHEIM, O.K. 1983. Aluminum speciation at the interface of an acid stream and a limed lake. Vatten 39: 301-304.

WRIGHT, R.F., GJESSING, E., CHRISTOPHERSEN, N., LOTSE, E., SEIP, H.M., SEMB, A., and SLETAUNE, B. in press. Project RAIN: changing acid deposition to whole catchments. The first year of treatment. Water Air Soil Pollut.

SOME ASPECTS OF THE CHEMICAL SPECIATION OF ALUMINIUM IN ACID SURFACE WATERS

E. TIPPING and C.A.BACKES

Freshwater Biological Association, Ambleside, Cumbria,
LA22 0LP, United Kingdom.

Summary

Aluminium in acid waters exists in a number of forms. The distribution of the metal among monomeric inorganic species can be calculated using established equilibrium constants. The determination and prediction of concentrations of organic monomeric Al and polymeric hydrolysed Al are more difficult. Here we discuss chemical kinetic analysis and cation exchange in measuring Al species, and the development of models that allow the calculation of organically complexed Al.

1. INTRODUCTION

In waters draining from acid soils, concentrations of Al are invariably high (> 100 µg/l). It is widely believed that some forms of the element are toxic to organisms, notably fish (see e.g. Baker, 1982). The primary objective of the present work was to investigate ways of distinguishing different chemical forms of Al. In addition, the possibility of predicting Al-organic complexation in natural waters was explored.

From an analytical point of view, it is convenient to divide Al in natural waters into three classes, viz. inorganic monomeric (Al^{3+} and its complexes with OH^-, F^- and SO_4^{2-}), organic monomeric (Al complexed by organic ligands) and colloidal or particulate Al (adsorbed Al, polymeric hydrolysis products, Al oxides and aluminosilicates). Concentrations of species in the inorganic monomeric fraction are interrelated by a series of equilibrium constants which appear to be reasonably well-established (see e.g. Burrows, 1977; May et al, 1979). Relatively little quantitative information is available on the complexation of Al with organic ligands in natural waters, although progress is being made (Young & Bache, 1985; Backes & Tipping, 1986/7 – see also below). Fieldworkers therefore estimate organic Al by separation techniques (e.g. Driscoll, 1984; LaZerte, 1984). Very little is known about colloidal and particulate forms of Al in natural waters: freshly formed $Al(OH)_3$ may be a mechanism of Al toxicity to fish (e.g. Baker, 1982). Colloidal and particulate Al could, via adsorption-desorption and precipitation-dissolution reactions, play a significant role in the control of Al speciation in streams and lakes.

Here, we report on the use of chemical kinetic analysis and cation-exchange separation in determining Al speciation, and on the development of models that relate the concentration of organically-complexed Al to the activities of Al^{3+} and H^+.

2. CHEMICAL KINETIC ANALYSIS

The principle of chemical kinetic analysis is that on the addition of excess of a suitable reagent, all the original forms of a chemical entity (Al in this case) will be converted to a single final form. Measurement of the rate of formation of the latter gives information on the types and amounts of the original forms. Smith (1971) showed that the technique could be applied to hydrolysed Al species. We attempted to extend the application to complexes of Al with F^- and humic substances, but were not successful because of the formation of intermediates - mixed complexes - during the analysis. We conclude that for determining Al speciation in natural waters, the use of chemical kinetic analysis is restricted to the identification and quantification of polymeric hydrolysed Al, and possibly other colloidal forms of the metal.

To perform kinetic analyses with Al we used a 'mixed reagent' consisting of hydroxyquinoline sulphonate (HQS) and NaF, buffered at pH 5.6. This reagent reacts quickly (< 1 minute) with momomeric forms of Al, including fluoride and humic complexes, to give a mixed HQS-Al-F complex, the colour of which is stable provided the temperature is kept constant. Other colour reagents for Al give complexes unstable with respect to time, and are suitable for kinetic analysis only in combination with another dissociating reaction, e.g. acid attack. The HQS/F reagent can be used to monitor continuously the dissociation of polymeric Al.

3. EVALUATION OF CATION-EXCHANGE FOR DETERMINING ORGANICALLY-COMPLEXED Al

The method of Driscoll (1984) is widely-used for estimating organically complexed Al. The principle of the method is that inorganic monomeric Al species are virtually all cationic and therefore adsorb to a cation exchange resin. Organic Al is assumed to be present as anions and therefore passes through the column. Measurement of total monomeric and monomeric Al after ion-exchange ('non-labile' Al in the terminology of Driscoll) allows inorganic monomeric Al to be estimated by difference. The method is convenient, reproducible and can be used in the field. However, application of the method requires that organically complexed Al in the original solution remains complexed during passage through the column.

We compared organic Al concentrations determined by cation exchange with those determined by equilibrium dialysis. The determinations were done on the same solutions, prepared from isolated humic substances (high molecular weight fraction) and $Al(NO_3)_3$ at pH 4 and 4.5. It was found that the cation exchange method underestimated organic Al by approximately 20%. Underestimation of organic monomeric Al necessarily causes an overestimation of inorganic monomeric Al. The seriousness of such overestimation depends on the proportion of Al in the organic form. An error equal to or greater than a factor of two occurs when ca. 80% or more of the Al is organically complexed. These findings indicate that results from the cation-exchange method should be interpreted cautiously, especially when they refer to organic-rich samples. The method nonetheless remains useful for obtaining an indication of the distribution of monomeric Al between organic and inorganic forms.

4. MODELLING Al-ORGANIC INTERACTIONS

The binding of Al by an isolated aquatic humic fraction was studied as a function of the activities of H^+ and Al^{3+}, by equilibrium-dialysis (Backes and Tipping 1986/7). Acid-base titrations were also carried out. Two models (I and II) of the $Al^{3+}-H^+$-humic interactions were formulated and tested with the experimental data.

Of the two models, I is simpler, and is based on the following equation:

$$\upsilon = \alpha(Al^{3+})\beta_{(H^+)}\gamma$$

where υ is the no. of moles of Al bound per g of humic substances, brackets -( ) - indicate activities, and α, β and γ are constants. This equation fitted the laboratory data tolerably well, with r = 0.93. For the humic fraction studied, the values of α, β and γ were 1.32×10^{-4} mol/gHS, 0.718 and -1.054 respectively.

Model II is based on the known properties of HS and emphasizes their polyelectrolyte nature. Interactions with charged species (H^+, Al^{3+}) are described in terms of intrinsic equilibrium constants and electrostatic attraction between negative charges on the humics and the positive charges of the complexed ions. The model assumes one class of n carboxylic acid groups, plus n/2 phenolic OH groups. Constants for model II were derived from acid-base titrations in the absence of Al, and from Al binding data obtained by equilibrium dialysis. The simplest form of model II (Backes and Tipping, 1986/7) involves five parameters: three intrinsic equilibrium constants, the number of binding sites, and an electrostatic interaction factor. For the humic fraction studied, this version of the model gives a fit with r = 0.93. Model II provides in addition to predictions of υ, estimates of the protons released per Al bound and of the net charge of the humic-Al complex.

The two models can be used to predict Al speciation in the presence of humic substances. With the present models, the predictions are rather imprecise; for example the 95% confidence limits for a predicted value of υ of 10^{-3}.mol/g HS (a middle-range value) are ca. $\pm 2.5 \times 10^{-4}$ mol/gHS. The models are most useful for predicting trends in Al binding with pH, (Al^{3+}) and humic concentration, and for assessing the magnitudes of competition effects. Further work with other humic fractions is required to test their generality. It is expected that higher precision will be obtained by increasing the complexity of model II, for example by incorporating heterogeneity of carboxylic acid groups, and allowing the electrostatic interaction factor to vary with humic charge.

Preliminary testing of the models with results for natural water samples has been encouraging. Model I was used to predict concentrations of organic Al in 36 field samples for which organic Al had been estimated by cation exchange (data from Johnson et al, 1981; Wright and Skogheim, 1983; Tipping, Woof and Ohnstad unpublished). Linear regression analysis gave the relationship [observed organic Al] $= 0.94 \times$ [calculated organic Al] $- 4 \times 10^{-7}$ (concentrations in mol l^{-1}) with r = 0.91.

5. CONCLUSIONS

On the basis of our studies, we consider that a reasonably satisfactory approach to the determination of the speciation of monomeric Al is that of Driscoll (1984), i.e. the estimation of organic Al by cation exchange separation, the estimation of inorganic Al by difference, and the calculation of inorganic speciation using established equilibrium constants. From our investigation of the performance of the cation-exchange method, however, we urge caution in applying it to samples with a high proportion of the Al in an organically complexed form. It might well be advisable to apply a correction of 20% to estimates of organic Al (see section 3).

Kinetic analysis, using either HQS/F or acid dissolution coupled with an appropriate analytical technique for monomeric Al, offers promise for the detection and quantification of polymeric hydrolysis products of Al.

The application of Al-organic complexation models appears to be useful in checking determinations of Al speciation and in predicting trends in speciation.

6. ACKNOWLEDGEMENTS

We thank J. Hilton and R. Harriman for helpful discussions, M. Ohnstad and C. Woof for technical assistance, and J. Hawksford for typing the manuscript.

REFERENCES

BACKES, C.A. and TIPPING, E. (1986/7). Water Research, in press.
BAKER, J.P. (1982). in Acid Rain/Fisheries (R.E. Johnson, ed.) (Amer. Fish. Soc., Bethesda) pp. 156-176.
BURROWS, W.D. (1977). Crit. Rev. Environ. Control. 7 167.
JOHNSON, N.H., DRISCOLL, C.T., EATON, J.S., LIKENS, G.E. and McDOWELL, W.H. (1981). Geochim. Cosmochim. Acta. 45 1421.
LAZERTE, B.D. (1984). Can. J. Fish. Aquat. Sci. 41 766.
MAY, H.M., HELMKE, P.A. and JACKSON, M.L. (1979). Geochim. Cosmochim. Acta. 43 861.
SMITH, R.W. (1971). Adv. Chem. 106 250.
WRIGHT, R.F. and SKOGHEIM, O.K. (1983). Vatten 39 301.
YOUNG, S.D. and BACHE, B.W. (1985). J. Soil Sci. 36 261.

EXPOSURE OF SMALL-SCALE AQUATIC SYSTEMS TO VARIOUS DEPOSITION LEVELS OF AMMONIUM, SULPHATE AND ACID RAIN

J.A.A.R. Schuurkes
Laboratory of Aquatic Ecology, Catholic University,
Toernooiveld, 6525 ED Nijmegen, The Netherlands

SUMMARY

In a greenhouse two experiments were conducted in which small-scale aquatic ecosystems were exposed to different synthetic rain solutions. Both rain containing sulphuric acid and rain containing ammonium sulphate were used to create different deposition levels of acid, sulphur and ammonium-nitrogen. In the first experiment seven identical soft water systems were exposed during a two year period to rain differing in pH (5.6 to 3.5), ammonium (1.4 to 8.5 kmol/ha/year) and sulphate (0.1 to 4.4 kmol/ha/year). From the acidic sulphate treatments only the pH=3.5 treatment caused a decrease in pH of the water (from 5.5 down to 4.2). Within the same period the pH=5.6 ammonium sulphate treatments caused a more rapid and stronger acidification (down to pH 4.1-3.5). Dose-effect relations were established between deposition of acid, nitrogen and sulphur on the one hand, and acidification, waterquality and vegetation on the other hand. Deposition containing high amounts of ammonium sulphate may be of quantitative importance in regulating water acidification as a result of the nitrification process. Vegetation development was mostly related to the level of nitrogen deposition and acidification of the water. In the second experiment two acid water systems were exposed to N- and S-polluted and unpolluted rain. During the first year the unpolluted rain resulted in an increase in pH (from 4.6 to 6.4) and alkalinity (plus 50 µeq/l.) of the water layer. A remarkable phenomenon in one of these systems was the appearance of Lobelia dortmanna, which had disappeared from its natural growing-site after 1957. Reversibility of acidification in acid stressed systems appeared to be possible under certain conditions.

1. INTRODUCTION

In The Netherlands several thousands of small water bodies are situated on the pleistocene sandy soils in the southern, middle and eastern parts of the country and in the northern coastal dune area. Both the water bodies and their catchment areas exhibit a very low buffering capacity and are poor in nutrients. Particularly the lentic and hydrologically isolated soft waters (small lakes, moorland pools, ponds and dune pools) are fully dependent on rain water and are very susceptible to acidifying atmospheric deposition. Nowadays about 90% of these waters are acid and 35% of them have a pH-value below 4.0. The alkalinity of 70% of the waters is less than 0.1 meq./l (Leuven and Schuurkes, 1985). Based on time trends in water quality data, temporal shifts in aquatic biota, and application of current empirical models for lake acidification, it appears that 59 - 96% of the Dutch soft waters have become acidified recently (Leuven et al., 1986a; Schuurkes and Leuven, 1986). Several experimental approaches have shown, that in regions susceptible to acid precipitation the present deposition, particularly of

NH_x and SO_2, causes a strong and rapid acidification of the water and lead remarkable changes in vegetation, similar to the altered conditions of many acidified waters (Roelofs, 1986; Schuurkes, 1986). Nitrification of air-borne ammonium sulphate appeared to play a dominant role in the acidification process. This is a rather neglected field of interest as in the acidification problem most attention is given to the impact of precipitation containing sulphuric and nitric acids. However, ammonium is a recognized component of acid precipitation and recently its importance in the acidification of susceptible soils and waters is gaining more attention (van Breemen et al., 1982; Anonymous, 1983; Galloway and Dillon, 1983; van Aalst, 1984; Fuhrer, 1985; Mosello and Tartari, 1985; Nihlgård, 1985; Roelofs et al., 1985; Roelofs, 1986). Particularly in agricultural and stockbreeding areas ammonium sulphate deposition is enhanced as a result of gaseous NH_3 emitted from animal manure and breeding farms. A recent survey of NH_3-emission in Europe (Buysman et al., 1985) shows that this atmospheric pollutant is of international importance. In many parts of The Netherlands, Belgium, the Federal Republic of Germany, Denmark, France, Great Britain and Italy ammonia-emission from non-industrial but biological sources is remarkably high. Acidification of surface waters has been established in most of these countries (Rebsdorf, 1980; Schoen et al., 1984; Vangenechten et al., 1984; Mosello and Tartari, 1985; Wieting, 1985). These observations may stress the need for an international study on effects of atmospheric ammonia deposition with respect to water acidification.

In the present contribution attention is focused on effects on soft and acidified aquatic systems caused by changes in atmospheric deposition of ammonium-nitrogen, sulphate and acid rain. The set up and results of two experimental studies are outlined. One study concerns experimental acidification by various rain treatments containing ammonium sulphate, nitrate and sulphuric acid. In order to study reversibility of acidification two acidified waters were exposed to both ammonium sulphate polluted and unpolluted rain. Dose-effect relations are established between nitrogen, sulphur and acid on the one hand, and acidification, alkalinization, water quality and vegetation on the other hand.

2. ACIDIFICATION BY VARIOUS DEPOSITION LEVELS OF AMMONIUM-NITROGEN, SULPHATE AND ACID RAIN.

In a greenhouse seven identical small-scale ecosystems were created, simulating field conditions of hydrologically isolated lentic soft waters. Mineral sandy sediment, poor in carbonate, was used and in all systems several macrophyte species from soft and acid waters were introduced in the same quantity and arrangement. During a two year period the systems were exposed to various artificial rain solutions differing in pH and ammonium, sulphate and nitrate concentrations. Two staged deposition gradients were created. One gradient was formed by treatments 1, 2, 3 and 4, being a sulphuric acid, nitrate containing pH-gradient. The other consisted of treatments 1, 5, 6 and 7 and represented a slightly acid ammonium sulphate gradient. The corresponding deposition data of acid, nitrogen and sulphur are given in table 1. The applications 2 to 6 are based on realistic deposition values of atmospheric pollutants in The Netherlands, Treatment 1 served as a reference. A full description of the experimental set up and conditions is given by Schuurkes et al. (in prep.).

2.1 Water chemistry and processes

Time trends of pH, alkalinity, ammonium, nitrate and sulphate concentrations are shown in figure 1. During the first year water quality development was generally subjected to large fluctuations. The rain applications affect the water both directly and indirectly as a result of physical and chemical interactions with the sediment. Only during the second year a rather stable situation in water chemistry was achieved. Apparently an equilibrium was established between water quality, rain application, sediment and vegetation. Ultimate water quality characteristics (figure 2) are based on values measured during this period.

Table 1: Some chemical characteristics of the rain applications.

no.	pH	concentration ($\mu mol.l^{-1}$)				deposition ($kmol.ha^{-1}.yr^{-1}$)					deposition ($kg.ha^{-1}.yr^{-1}$)		
		H^+	NH_4^+	SO_4^{2-}	NO_3^-	H^+	NH_4^+	SO_4^{2-}	NO_3^-	H^+(pot.)	NH_4^+-N	NO_3^--N	SO_4^{2-}-S
1	5.6	2	15	15	20	0.01	0.08	0.07	0.08	0.17	1.1	1.1	2.2
2	5.0	10	15	15	45	0.05	0.07	0.06	0.22	0.19	1.0	3.1	1.9
3	4.25	50	15	55	45	0.25	0.06	0.23	0.22	0.37	0.8	3.1	7.4
4	3.5	320	15	210	45	1.55	0.07	0.97	0.20	1.69	1.0	2.8	31.0
5	5.6	2	265	160	2	0.01	1.38	0.82	0.01	2.77	19.3	0.1	26.2
6	5.6	2	510	300	2	0.01	2.87	1.61	0.01	5.75	40.2	0.1	51.5
7	5.6	2	1585	785	2	0.01	8.45	4.40	0.01	16.91	118.3	0.1	140.8

The development of water quality of the systems differs markedly between the various rain treatments. Water acidification occurs in system 4, 5, 6 and 7. Apparently ammonium sulphate addition has a strong acidifying effect. Two processes may be involved; one being nitrification of ammonium, which produces 2 moles hydrogen and 1 mol nitrate per mol converted ammonium. The other being ammonium uptake by macrophytes and algae which often is coupled with an equivalent release of hydrogen. The development of both ammonium and nitrate in the water layer indicates the importance of the nitrification process. At ammonium concentrations more than 10 μM. nitrifying bacteria are able to outcompete autotrophic organisms with respect to ammonium uptake (Brown and Johnson, 1977). Higher ammonium concentrations cause a more rapid acidification and result in lower pH-values. The quantitative importance of nitrification in the acidification process is shown in figure 4. A relation can be noticed between the pH of the water and the applied amount of potential acid (table 1), which was calculated as the sum of hydrogen and twice the amount of ammonium. Particularly after one year the degree of acidification (calculated from the decrease in alkalinity and increase in H^+-concentration) shows a linear correlation with the applied amount of potential acid (Schuurkes et al., 1986a). This leads to the conclusion that ammonium in precipitation has a potential acidifying capacity in soft waters and should be taken into account in the "acid rain" problem. Its role in the acidification process will be of quantitative importance in hydrologically isolated soft waters situated in NH_3 affected areas (Schuurkes, 1986). Low pH-values (below 5) seem to be an important limiting factor for the nitrification process in aquatic systems (Rao and Dutka, 1983). In the experiment described, the relative increase in H^+-concentration of the water in systems 5, 6 and 7 during the second year is low (figure 4). Nitrification apparently is inhibited at pH-values ≤4.25, while at pH=3.5 no nitrification takes place (figure 4). Below pH 3.7 a continuous ammonium-addition results in accumulation of ammonium (figures 1 and 2) which subsequently becomes available for acid-tolerant macrophyte species (see next paragraph).

The acid rain applications containing sulphuric acid (treatments 2, 3 and 4) affected pH and alkalinity most clearly during the second year. During the first year the acidifying capacity of the rain water apparently was not sufficient to decrease both the initial alkalinity and the possibly generated alkalinity. After two years of exposure only system 4 was strongly acidified. Due to the continuous addition of acid (pH = 3.5) rain, acidification occurred at the moment that bicarbonate alkalinity strongly decreased (figure 1).

37

In the systems where a strong acidification has occurred, the acidifying influence of rain water far outweighs any alkalinization process. However, such processes e.g. sulphate reduction and denitrification may be
involved but will be limited by the very low organic matter content of the
sediment (Knowles, 1982). Inhibition due to low pH-values will be of minor
importance (Postgate, 1979; Kelly and Rudd, 1984). The importance of the
alkalinizing role of nitrate uptake by macrophytes and algae cannot be
established but will certainly depend on the uptake capacity of the involved
species.

Rain containing nitrate (treatments 1 to 4) did not affect nitrate concentrations in the water. Nitrate concentrations are most clearly enhanced
during wintertime and spring as a result of the nitrification process. However, nitrate uptake by both autotrophic organisms and bacteria will be
responsible for the general relatively low concentrations in the water.

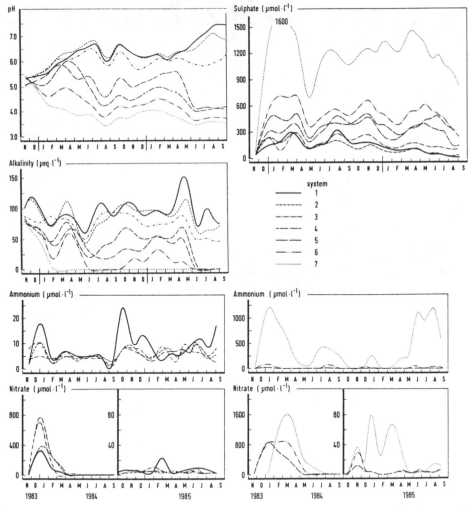

Figure 1: Development of pH, alkalinity, sulphate, ammonium and nitrate in
the water layer of seven small-scale systems exposed to different deposition
levels of acid, sulphur and nitrogen as given in table 1. (From: Schuurkes
et al., in prep.)

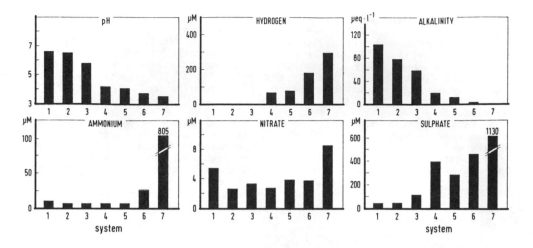

Figure 2: The ultimate levels of some chemical constituents in the water of systems 1 to 7 (mean values of the summer period).

The applied amount of sulphate (whether as sulphuric acid or ammonium sulphate) determines sulphate concentration in the water. Although concentrations may fluctuate the general level of the pathway is related to the applied amount of sulphate (figure 1). Sulphate concentrations in the water are higher than those of the corresponding rain additions. Apparently sulphate accumulates and sulphate reduction is of minor importance (figure 2). However, the sulphate concentrations primiraly depend on the sulphur richness of the sediment and sulphate weathering. This is of great importance for determining the excess of sulphate in acidifying surface waters.

2.2 Vegetation

The development and biomass of the aquatic and amphibic vegetation depends on the net result of deposition and water quality. Therefore the relation with the rain applications will be indirectly. In figure 3 the ultimate biomass of some aquatic macrophytes is shown. None of the sulphuric acid pH-treatments clearly affected the biomass of the introduced species. The most pronounced changes occurred in systems 6 and 7. Although the biomass of Littorella uniflora in these systems was much higher, it disappeared from the water layer. Pilularia globulifera completely disappeared. The biomass of the acid-tolerant species Juncus bulbosus, Sphagnum cuspidatum and Agrostis canina has increased in systems 5, 6 and 7. Particularly the ammonium sulphate treatments 6 and 7 clearly increased the biomass of these species.

The observed shift in macrophyte composition is the result of both acidification and ammonium enrichment. Submersed soft water species disappear, while acid-tolerant species profit from the enhanced N-availability. The described conditions as a result of the ammonium sulphate applications are of competitive advantage for these species. They are able to expand their area in the water layer when carbondioxide availibility is not limiting. However, increased carbondioxide-concentrations are often observed in acidifying waters. Roelofs et al. (1984), Wetzel et al. (1984) and Schuurkes et al. (1986b) describe the importance of this alteration in availibility of carbon and nitrogen. Acidification of the water layer itself does not directly affect macrophyte composition.

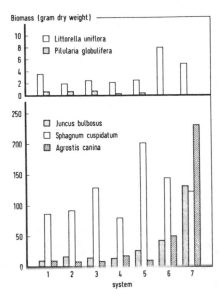

Figure 3: The biomass of some macrophyte species in seven systems after 22 months of exposure to different types of rain. (From: Schuurkes et al., in prep.)

2.3 General dose-effect relations
    The observed changes in water quality and vegetation of the simulated systems exposed to a high ammonium sulphate deposition correspond remarkably with the conditions of nowadays strongly acidified waters in NH3 affected areas. The pH of 35% of these waters is below 4.0 (mean pH=3.8) and both Sphagnum spp. and Juncus bulbosus often dominate the water layer. Under natural conditions nitrification of airborne ammonium may also account for the very low pH-values of acidified waters, being much lower than might be expected as a result from pH of precipitation. Therefore, atmospheric deposition containing high amounts of ammonium compounds is of quantitative importance in determining the rate and level of water acidification and has a pronounced effect on macrophyte composition. Deposition merely containing sulphuric and nitric acids also affect soft water systems but the time-related effects on water quality and vegetation are less "severe". Although in North America, Scandinavia and Canada these strong acids are the most important acidifying components, emphasis must be laid on the importance of ammonium deposition in regions where a relatively high emission of gaseous NH$_3$ occurs.
    The presented dose-effect relations indicate that only an increased deposition of acidifying components cause enhanced concentrations of H$^+$, sulphate and ammonium in the water layer (figures 1, 2 and 4). A continuous input of the same magnitude results in a rather stable situation in which only minor alterations occur. Several empirical models are available to estimate the degree of acidification of surface waters (Henriksen, 1982; Kramer and Tessier, 1982). These models generally are based on simple chemical approaches, which are deduced from the correlation between pH and excess of sulphate in Norwegian lakes on the one hand, and the excess of sulphate and H$^+$-concentration in precipitation on the other hand. Although several assumptions should be made before application (Kramer and Tessier, 1982), the empirical model of Henriksen gave good predictions for the level of acidification of susceptible waters in different parts of the world (Overrein et al., 1981). For Dutch soft waters the degree of acidification

was strongly correlated with the excess of sulphate (Leuven et al., 1986a). In the case of the presented experiment a linear relation was established between sulphate deposition and the sulphate concentration in the water (figure 4). The basic sulphate concentration, however, will depend on sedimentary conditions and physical processes as sulphate weathering. Whether only sulphuric acid in precipitation may be the major contributor to water acidification in general must be queried. Other chemical processes are also of importance, depending on the chemical composition of atmospheric deposition. As described before, nitrification of airborne ammonium may be of quantitative importance in NH_3 affected regions.

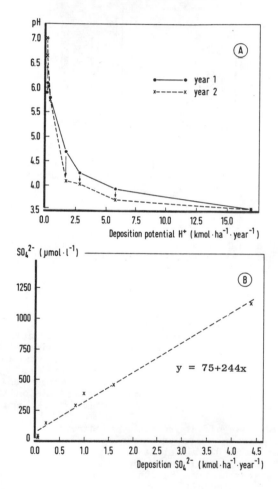

Figure 4: Relation between deposition of potential acid and pH (A); and deposition of sulphate and sulphate concentration (B). (From: Schuurkes et al., in prep.)

Another point of importance of the presented dose-effect relations is the use for the assessment of limit values for deposition of air pollutants. However, some restrictions must be made concerning extrapolation to natural waters. Based on these results, Schuurkes and Leuven (1986) determined limit values for deposition of acid, nitrogen and sulfur, being 250 mol H^+, 1400 mol N and 230 mol SO_4^{2-} yearly per ha. respectively.

3. REVERSIBILITY OF ACIDIFICATION IN ACID STRESSED SYSTEMS

Reduction of atmospheric deposition of acidifying substances is an important measure to restrict the continuing acidification process. However, a positive development towards the original soft-water ecosystems can only be expected when the acidifying processes are reversible. Primarily pH and alkalinity should be increased. An important question is whether the former-ly present flora and fauna species are able to re-establish or to expand their area in the restored soft waters. Problems may occur with respect to the water quality, as organic matter and nutrients have been accumulated during the acidification process (Grahn et al., 1974; Traaen, 1980; Hendrey, 1982). When pH and alkalinity increase, microbial decomposition will be enhanced and a release of accumulated nutrients from the sediment may be expected. Such difficulties influence the effectiviness of natural restoration of surface waters. However, in models predicting the effectiviness of reduced acidifying deposition in Scandinavian lakes (Wright and Henriksen, 1983), hardly any account is taken into these problems.

In the present paper the reversibility of acidification in Dutch waters as a consequence of reduced N- and S-deposition was studied by means of an experimental approach very similar to that described in the previous chapter. Emphasis was laid on both water quality development and the possible revival of soft water macrophytes. Mineral sediment was taken from two recently acidified waters (I and II) and a low-alkaline water (III). Some chemical characteristics of these waters are given in table 2. Each of the sediments was used to create two small-scale systems, comparable to ponds, simulating field conditions of hydrologically isolated lentic waters.

Table 2: Some chemical characteristics of the waters from which sediment was obtained (mean 1983 values in μM.; From: Leuven et al., 1986b).

		pH	alk.[*]	SO_4^{2-}	NH_4^+	NO_3^-	PO_4^{3-}	Al^{3+}	Cd^{2+}[**]
I	Peetersven	3.9	0	80	45	5	0.2	2.3	7.0
II	Groot Huisven	4.3	0.05	224	113	7	0.3	10.4	2.6
III	Beuven	6.9	0.62	479	24	1	1.7	6.5	4.9

[*] alk. = alkalinity in $meq.l^{-1}$

[**] expressed as $nmol.l^{-1}$

Each of the small-scale systems was exposed to a different synthetic rain solution (A and B: table 3). Rain treatment A simulated slightly acid rain and was based on atmospheric deposition of nitrogen and sulphur in rather unpolluted regions. Treatment B was based on deposition of acid, nitrogen and sulphur in NH_3 affected areas in The Netherlands (Schuurkes, 1986).
Water quality of the systems was monitored every two weeks whereas vegetation development was registered only at the end of the study period. In this contribution only some preliminary results of the first six months are described. Meanwhile the experiment is continued for another year.

Table 3: Some chemical characteristics of the synthetic rain applications A and B.

	pH	H^+	NH_4^+	SO_4^{2-}	NO_3^-	H^+	NH_4^+	SO_4^{2-}	NO_3^-
		concentration(μmol.l^{-1})				deposition(kmol.ha^{-1}.$year^{-1}$)			
A	5.6	2	5	15	40	0.01	0.02	0.06	0.17
B	5.6	2	550	330	2	0.01	2.36	1.42	0.01

Time trends of pH and alkalinity are shown in figure 5. The effects of treatment A and B on water quality were remarkably different. The polluted rain (treatment B) caused a rapid and strong acidification (pH $\leq$4) of the water in the systems. Here, nitrification of ammonium is of quantitative importance in regulating the acidification process (see previous chapter). The effect of the unpolluted rain on water quality is different for the acid systems (I and II) and the circum neutral reference system (III). After some months pH and alkalinity of system I and II increased, whereas both parameters were rather constant in system III. Apparently in both acid systems a net alkalinity production occurred as a result of the very low deposition of acid, sulphate and nitrogen and the subsequent physical and chemical processes. The originally present acid water is slowly replaced by less acidic rain. In the meantime alkalinity is generated by silicate weathering and probably as a result of sulphate- and nitrate reduction or other alkalinizing processes. A decreased input of acidifying substances (the unpolluted rain treatments) resulted in lower concentrations of sulphate, nitrogen, calcium, magnesium, aluminium and cadmium in the water layer of the systems (table 4).
Effects on vegetation development during the first growing season were generally less clear, although some remarkable phenomena were observed. Particularly the appearance of Lobelia dortmanna in system I exposed to unpolluted rain was of special importance as this species had disappeared from its natural growing-site after 1957. Apparently, the more favourable chemical conditions of water and sediment enables Lobelia to develop well after germination.

Table 4: Concentration of some ions* in the water of three different systems exposed to unpolluted (A) and polluted rain (B) as given in table 3.

system		I		II		III	
treatment		A	B	A	B	A	B
NH_4^+	μmol/l.	8	253	5	424	11	111
NO_3^-	"	8	11	7	12	4	15
SO_4^{2-}	"	23	374	30	334	83	415
Ca^{2+}	"	8	43	11	20	66	183
Mg^{2+}	"	14	40	20	30	31	71
Al^{3+}	"	12	56	5	26	6	14
Cd^{2+}	ηmol/l.	2	3	2	2	2	5

* mean values from July- September

43

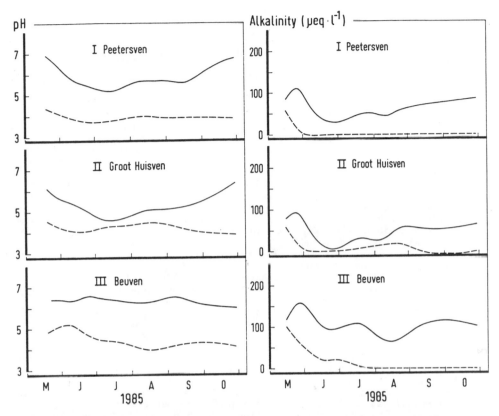

Figure 5: Development of pH and alkalinity in the water layer of three systems exposed to rain applications A (——) and B (----).

The results of the presented experimental approach indicate that
reversibility of natural aquatic ecosystems as a result of a severe reduction of atmospheric deposition of acid, sulphur and nitrogen is possible.
However, the experiment concerns simulated systems with a sediment poor in
organic matter and assumptions must be made before extrapolation of the
results to natural waters. Also generalizations are of limited value as the
geological and hydrological conditions of susceptible waters in The Netherlands are different from e.g. those of Scandinavian lakes. Nevertheless, the
research approach and results will be applicable for susceptible waters in
many other western-European countries as pointed out in the introduction.
However, a more extensive set up is needed with respect to different conditions of both atmospheric deposition and sediment susceptibility. Such an
approach allows to establish internationally applicable empirical models for
predicting effectiviness of various deposition scenarios. In addition,
attention should be focused on the possibilities of the disappearing flora
and fauna species to re-establish and/or expand in the restored waters, as
well as on the role of accumulation of organic matter in reducing the effectiviness of water quality recovery.

ACKNOWLEDGEMENTS
    The publication of the results is sponsored by the Ministry of Housing,
Physical Planning and Environment, Directorate Water. The author is greatly
indebted to Prof. Dr. C. den Hartog, Dr. B. Higler, Drs. R.S.E.W. Leuven and
J.G.M. Roelofs for critically reading the manuscript.

REFERENCES

1. AALST, R.M., VAN (1984). Verzuring door atmosferische depositie - Atmosferische processen en depositie. Ministery of Environment. Publication series "Milieubeheer" no. 84.
2. ALEXANDER, M. (1977). Introduction to soil microbiology. Wiley, New York.
3. ANONYMOUS, (1983). Acid rain, a review of the phenomenon in the EEC and Europe. Ed. Environmental Resources Ltd. Graham and Trotman. London.
4. BREEMEN, N. VAN, BURROUGH P.A., VELTHORST E.J., DOBBEN H.F., DE WIT T., DE RIDDER,T.B. and REYNDERS H.F. (1982): Soil acidification from atmospheric ammonium sulphate in canopy throughfall. Nature 299: 548-550.
5. BROWN, C.M. and JOHNSON B. (1977). Inorganic nitrogen assimilation in aquatic micro-organisms. Adv. Aquat. Microbiol. 2: 49-101.
6. BUYSMAN, E., MAAS, J.F.M. and ASMAN, W.H.M. (1985). Ammonia emission in Europe. Report R-85-2. Institute for Meteorology and Oceanography, Rijksuniversiteit Utrecht, 28 pp.
7. FUHRER, J. (1985). Formation of secondary airpollutants and their occurrence in Europe. Experientia 41: 286-301.
8. GALLOWAY, J.N. and DILLON, P.J. (1983). Effects of acid deposition: the importance of nitrogen. In: Ecological effects of acid deposition. National Swedish E.P. Board - Report PM 1636.
9. GRAHN, O., HULTBERG, H. and LANDNER, L. (1974): Oligotrophication: a self accelerating process in lakes subjected to excessive supply of acid substances. Ambio 3: 93-94.
10. HENDREY, G.R. (1982): Effects of acidification on aquatic primary producers and decomposers. In: Acid Precip. Fish. Imp.: 125-135. R.E. Johnson (Ed.). Amer. Fish. Society, Bethesda, Maryland.
11. HENRIKSEN, A. (1982). Susceptibility of surface waters to acidification. In: Acid Rain/Fisheries. Proc. Int. Symp. Acid. Precip. Fish. Imp.: 103-121. Ed. R.E. Johnson. American Fisheries Society, Bethesda, Maryland.
12. KELLY, C.A. and RUDD, J.W.M. (1984). Epilimnic sulphate reduction and its relationship to lake acidification. Biogeochemistry 1: 63-67.
13. KNOWLES, R. (1982). Denitrification. Microbiol. Rev., 46: 43- 70.
14. KRAMER, J. and TESSIER, A. (1982). Acidification of aquatic ecosystems: a critique of chemical approaches. Environ. Sci. Technol., 16: 606-615.
15. LEUVEN, R.S.E.W. and SCHUURKES, J.A.A.R. (1985). Effecten van zure neerslag op zwak gebufferde en voedselarme wateren. Ministery of Environment, Publication series "Lucht" no. 47.
16. LEUVEN, R.S.E.W., KERSTEN, H.L.M., SCHUURKES, J.A.A.R., ROELOFS, J.G.M. and ARTS, G.H.P. (1986a). Evidence for recent acidification of lentic soft waters in The Netherlands. Water, Air and Soil Pollution (in press).
17. LEUVEN, R.S.E.W., VAN DER VELDEN, J.A., VANHEMELRIJK, J.A.M. and VAN DER VELDE, G. (1986b): Impact of acidification on chironomid communities in poorly buffered waters in The Netherlands. Ent. Scand. Suppl. (in press).
18. MOSELLO, R. and TARTARI, G. (1982): Effects of acid precipitation on subalpine and alpine lakes. Water Quality Bulletin.
19. NIHLGARD, B. (1985). The ammonium hypothesis: an additional explanation to the forest die-back in Europe. Ambio 14(2).
20. OVERREIN, L.N., SEIP, H.M. and TOLLAN, A. (1981). Acid precipitation - effects on forests and fish. Final report SNSF-project 1972-80. Oslo-Aas.
21. POSTGATE, J.R. (1979). The sulphate reducing bacteria. University Press, Cambridge.
22. RAO, S.S. and DUTKA, B.J. (1983). Influence of acid precipitation on bacterial populations in lakes. Hydrobiologia, 98: 153-157.
23. REBSDORF, A. (1980). Acidification of Danish soft waters. In: Proc. Int. Conf. Ecol. Imp. Acid. Precip.: 238-239. Eds. D. Drabløs & A. Tollan. SNSF project. Oslo-Aas.

24. ROELOFS, J.G.M., SCHUURKES, J.A.A.R. and SMITS, A.J.M. (1984). Impact of acidification and eutrophication on macrophyte communities of soft waters in the Netherlands. II Experimental Studies. Aquat. Bot. 18: 389-411.
25. ROELOFS, J.G.M., KEMPERS, A.J., HOUDIJK, A.L.F.M. and JANSEN, J. (1985). The effect of airborne ammonium sulphate on Pinus nigra var. maritima in The Netherlands. Plant and Soil, 84: 45-56.
26. ROELOFS, J.G.M. (1986). The effect of airborne sulphur and nitrogen deposition on aquatic and terrestrial heathland vegetation. Experientia, 42: 372-377.
27. SCHOEN, R., WRIGHT, R. and KRIETER, M. (1984). Gewässerversauerung in der Bundesrepublik Deutschland. Naturwissenschaften 71: 95-97.
28. SCHUURKES, J.A.A.R. (1986). Atmospheric ammonium sulphate deposition and its role in the acidification and nitrogen enrichment of poorly buffered aquatic systems. Experientia 42: 351-357.
29. SCHUURKES, J.A.A.R., HECK, I.C.C., HESEN, P.G.M., LEUVEN, R.S.E.W. and ROELOFS, J.G.M. (1986a). Effects of sulphuric acid and acidifying nitrogen deposition on water quality and vegetation of simulated soft water ecosystems. Water, Air and Soil Pollution, accepted.
30. SCHUURKES, J.A.A.R., KOK, C.J. and DEN HARTOG, C. (1986b). Ammonium and nitrate uptake by aquatic plants from poorly buffered and acidified waters. Aquat. Bot. 24(2): 131-147.
31. SCHUURKES, J.A.A.R. and LEUVEN, R.S.E.W. (1986). Verzuring van oppervlaktewateren door atmosferische depositie: oorzaken, omvang en effecten. Ministery of Environment. Publication series "Lucht" (in press).
32. SCHUURKES, J.A.A.R., ELBERS, M.A., GUDDEN, J.J.F. and ROELOFS, J.G.M. Effects of ammonium, sulphate and acid rain on acidification, water quality and vegetation of small-scale soft water systems. Aquat. Bot. (in prep).
33. TRAAEN, T.S. (1980): Effects of acidity on decomposition of organic matter in aquatic environments. Proc. Int. Conf. Ecol. Imp. Acid. Precip.: 340-341. D. Drabløs and A. Tollan (Eds.). Norway, SNSF project.
34. VANGENECHTEN, J.H.D., VAN PUYMBROECK, S. and VANDERBORGHT, O.L.J. (1984). Acidification in Campine bog lakes. In: Acid deposition and the sulphur cycle. Proc. Symp.: 251-262. Ed. O.L.J. Vanderborght. SCOPE, Belgium.
35. WETZEL, R.G., BRAMMER, S.E. and FORSBERG, C. (1984). Photosynthesis of submersed macrophytes in acidified lakes. I. Carbonfluxes and recycling of CO2 Juncus bulbosus (L.). Aquat. Bot., 19: 329-342.
36. WIETING, J. (1985). Water acidification by air pollutants in the Federal Republic of Germany. In: Abstracts Muskoka Conference Canada: 81-82.

THE CHEMICAL AND BIOLOGICAL FEATURES OF POORLY BUFFERED IRISH LAKES

J. J. BOWMAN

An Foras Forbartha

DUBLIN

A significant number of small Irish lakes have poorly buffered waters and thus are sensitive to artificially acidified rain. The chemistry and biology of four such lakes, alkalinity range 4 - 7 mg $CaCO_3/\ell$, two near the east coast and two in the west, have been studied over a two year period. The chemistry of the east coast lakes is influenced by atmospheric inputs of H^+ and SO_4 and by humic acids, whereas the west coast lakes appear to be largely influenced by natural inputs of acidity.

One lake in each area (Maumwee Lough in the west and Glendalough lake in the east) was examined biologically. 231 planktonic algal taxa, mostly Chlorophyceae, were recorded in Maumwee Lough compared to 108 in Glendalough Lake. A greater diversity of zooplankton was noted in the west coast lake. The littoral macroinvertebrate population in Maumwee Lough, dominated by Caenis moesta, was consistent with that of a lake unimpacted by acidity while that of the east coast lake was typical of a lake under acid stress. Salmon, trout, minnow and eels were present in the Maumwee system while trout and minnow only were taken in a survey of fish stocks in the Glendalough system.

It is suggested that the biological and chemical characteristics of Maumwee Lough would provide a baseline for assessing the impact of acid rain on other areas.

INTRODUCTION

The results of measurements in recent years indicate that precipitation in many areas of Europe and North America is artificially acidified, and that the pollutants causing this have been transported, in the atmosphere, long distances from the industrial centres to more remote areas. The increased acidity of the precipitation has been associated with the acidification of sensitive aquatic ecosystems downwind of major emission areas, resulting in serious ecological problems in lakes and rivers. In Ireland the prevailing wind is westerly and the precipitation associated with air masses originating in the Atlantic Ocean is not artificially acidified. However, precipitation borne by easterly winds from Britain and the European landmass tends to be more acidic. The purpose of the investigation described below was to determine the impact, if any, of acid rain on four selected Irish lakes of naturally poor buffering capacity viz. Loughs Maumwee and Nahasleam in Co. Galway (West of Ireland) and Lough Bray Lower and Glendalough Lake Upper in Co. Wicklow (East of Ireland) and to provide a baseline against which similar water

47

TABLE 1. Summary of the Morphological and Topographical Characteristics of Loughs Maumwee, Nahasleam, Bray Lower and Glendalough Lake Upper.

Lake	National Grid Ref.	Altitude M.O.D.*	Catchment km²	Av. Ann. Rain mm a⁻¹	Evapotranspiration losses mm a⁻¹	Annual Outflow m³ sec⁻¹	Surface Area km²	Volume m³ 10⁺⁶	Av. Depth m	Max. Depth m	Water Renewal rate a
Co. Galway (West Coast)											
Lough Maumwee	L 977 485	47.9	4.3	2150	462	0.23	0.27	0.610	2.3	5.4	0.084
Lough Nahasleam	L 972 440	34.11	24.1	1971	462	1.15	0.29	0.382	1.3	6.15	0.011
Co. Wicklow (East Coast)											
Lough Bray Lower	O 136 162	374.4	1.28	1750	423	0.054	0.29	5.857	20.2	47.6	3.5
Glendalough Lake Upper	T 102 962	133.5	18.7	2078	423	0.98	0.38	6.369	16.8	33.2	0.21

*Poolbeg Datum.

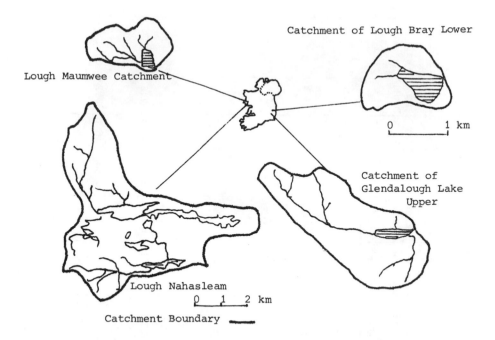

Fig. 1. Map showing catchments of four lakes (hatched) sampled and
their feeder streams.

bodies in other E.E.C. member states, the geographical position of which
makes them more exposed to artificially acidified rain, can be compared.
The investigations were supported by the E.E.C. under the Research and
Development Programme in the field of Environment (Contract No. ENV 784
(IRL)).

The following brief summaries outline the main features of the four
lakes and their catchments (see Table 1 and Fig. 1) while a more detailed
description of the morphological and topographical characteristics of
these lakes and of the sampling and analysis carried out is given by
Bowman (1986).

Lough Maumwee is a shallow lake whose catchment lies between 48 and
525 m over datum. The principal rock in the catchment is a Quartzite
Formation while the area immediately around the lake is on Granite. The
principal soils are Lithosols with outcropping rock (70%) and are
associated with blanket peat (25%) and peaty podzols (5%).

Lough Nahasleam is a shallow, rocky lake whose catchment lies between
40 and 300 m over datum. The principal rock in the catchment is
Paragneiss derived from members of the Dalradian sequence and is variably
Hornfelsic and Migmatitic. A considerable area in the northern portion
of the catchment lies on marble.

Lough Bray Lower is a relatively deep lake whose catchment lies
between 350 and 500 m over datum. The lake is fed by two small streams
entering on the western shore. The catchment and lake lie on granite

TABLE 2. The weighted mean pH value, and the weighted mean concentrations (μ eq ℓ^{-1}) of H^+ "excess" SO_4, Ammonia, Nitrate, Nitrite and Sodium in the accumulated rainfall samples in Counties Galway and Wicklow between April 1984 and December 1985.

		pH	H^+	"Excess" Sulphate	Ammonia	Nitrate	Nitrite	Sodium
Co. Wicklow	1984	4.92	12	30	33	21	0.6	70
Lough Bray Lower	1985	4.22	67	48	43	38	0.4	129
	1984	5.05	9	40	34	20	0.8	170
Glendalough Lake Upper	1985	4.24	57	57	29	33	0.3	230
Co. Galway	1984	4.58	26	30	9	15	0.2	350
Lough Maumwee	1985	4.37	42	24	10	18	0.2	330
	1984	5.1	8	10	5	6	0.3	330
Lough Nahasleam	1985	4.38	42	47	10	18	0.2	374

bedrock overlain by Peaty Podzols (75%) associated with Blanket Peat and Lithosols (25%.

The Upper Lake at Glendalough is a relatively deep lake whose catchment lies between 120 and 700 m over datum. The greater part of the catchment lies on Granite bedrock while the area immediately surrounding the lake is on Ordovician rock. The dominant soil type in the catchment is Lithosols with outcropping rock (70%) in association with Blanket Peat (25%) and Peaty Podzols (5%).

The physicochemical characteristics of all four lakes and their feeder streams were studied over the two year period of the investigation commencing in January 1984; biological studies (plankton, fish stocks, benthic flora and fauna) were confined to two of the lakes viz. Lough Maumwee and Glendalough Lake Upper. Standard collectors were installed at appropriate points in each catchment to obtain samples of precipitation (total).

PRECIPITATION

Fisher (1982) analysed monthly measurements of airborne and precipitated sulphur over Ireland to determine the annual wet and dry deposition. It was suggested that 50% of the airborne sulphur came from outside the state, chiefly from Britain. It has been calculated that in 1983 the emissions of SO_2 and NO_x within Ireland, based on fuel consumption, were 140,000 and 57,000 tonnes, respectively (An Foras Forbartha, 1984). The principal sources of sulphate in 1983 were the industrial and commercial sector (49%) and power stations (31%), while the main contributors to the NO_x emissions were power stations (33%), transport (33%) and industry (26%) (Bailey and Dowding, 1985).

The results of analyses of precipitation from the locations on the east and west coasts of Ireland collected during this investigation are set out in Table 2. Concentrations of "excess" sulphate were higher in the precipitation on the east coast than on the west with a marked increase in the levels between 1984 and 1985. A similar situation was also noted for nitrate. Ammonia concentrations were particularly low at the Loughs Maumwee and Nahasleam sampling points in the west of Ireland, possibly reflecting the low level of agricultural activity in those areas; low ammonia levels may be a factor contributing to the low pH values recorded at these western locations, particularly at Lough Maumwee. In contrast, the weighted mean H^+ concentrations at a site with greater agricultural activity 6 km north west of Maumwee Lough, based on daily samples, was 11 and 32 μ eq ℓ^{-1} for 1984 and 1985, respectively. A marked increase in the H^+ concentration between 1984 and 1985 was noted at all stations, and maximum concentrations greater than 150 μ eq ℓ^{-1} were recorded in the precipitation at the east coast locations in 1985.

THE LAKES AND FEEDER STREAMS

Chemical Characteristics

The arithmetic mean values for the standard limnological parameters in Loughs Maumwee, Nahasleam, Bray Lower and Glendalough Lake Upper are given in Table 3. The individual values recorded for pH, colour, total and "non-marine" sulphate, and total aluminium for each lake (1 m depth samples) and the principal inflowing stream to each lake on each sampling date are shown in Figs. 2 - 5.

The contents of dissolved ionic solids, as indicated by conductivity, were higher in the waters of the west coast lakes compared to those on the east coast. The greater precipitation in 1985 may account for the lower conductivities recorded in the lakes in that year. The mean pH values

TABLE 3. The annual mean concentrations of the principal limnological parameters (at 1 m depth) in Loughs Maumwee, Nahasleam, Bray Lower and Glendalough Lake Upper, 1984 – 1985.

	West Coast Lakes				East Coast Lakes			
	L. Maumwee		L. Nahasleam		L. Bray Lower		Glendalough L. Upper	
	1984	1985	1984	1985	1984	1985	1984	1985
Conductivity μS cm	90	50	105	65	45	45	45	40
pH	6.00	5.74	6.36	5.98	4.69	4.33	5.94	5.31
Colour Hazen	20	30	25	55	35	45	20	40
Nitrate mg N m^{-3}	23	68	69	99	97	139	356	363
Nitrite mg N m^{-3}	2	4	4	3	3	5	3	4
Ammonia mg N m^{-3}	20	18	24	23	32	48	29	29
Total Phos. mg P m^{-3}	9	12	8	11	17	22	8	15
Orthophosphate mg P m^{-3}	1	2	1	1	5	8	1	2
Alkalinity mg CaCO$_3$ ℓ^{-1}	5	6	7	7	4	3	5	5
Sulphate mg SO$_4$ ℓ^{-1}	4.64	2.94	5.21	3.68	5.25	5.20	4.84	4.0
Non-Marine Sulphate mg SO$_4$ ℓ^{-1}	1.75	1.55	1.80	2.07	4.04	4.18	3.64	3.13
Chlorophyll 'a' mg Chl $\underline{a}$ m^{-3}	1.6	1.8	1.4	1.4	12.0	4.0	0.80	0.92
Chloride mg Cℓ ℓ^{-1}	23	11	27	15	9	8	8	7

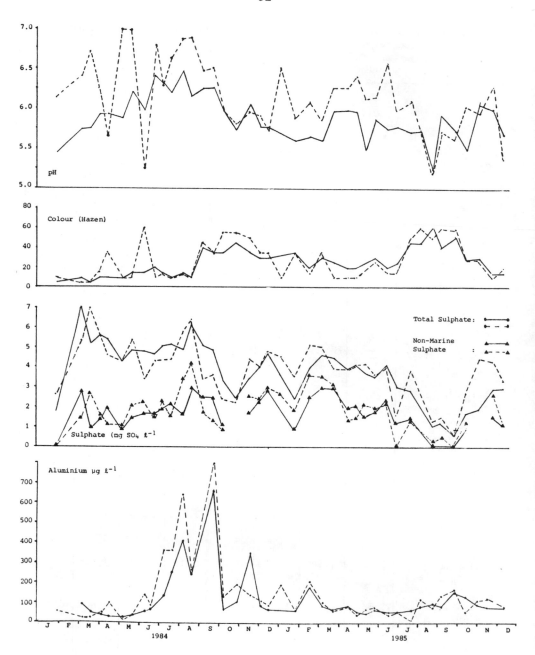

Fig. 2.   Lough Maumwee and its inflowing stream.   Variation in the
concentrations recorded in the lake (1 m dpeth) and in the
inflowing stream for pH, colour, sulphate and aluminium in
1984 and 1985.

Lake (1 m depth) ————— ;   Inflowing Stream -----

Fig. 3.   Lough Nahasleam and its inflowing stream.   Variation in the
           concentrations recorded in the lake   (1 m depth) and in the
           inflowing stream for pH, colour, sulphate and aluminium in
           1984 and 1985.

           Lake (1 m depth)————— ;   Inflowing ----

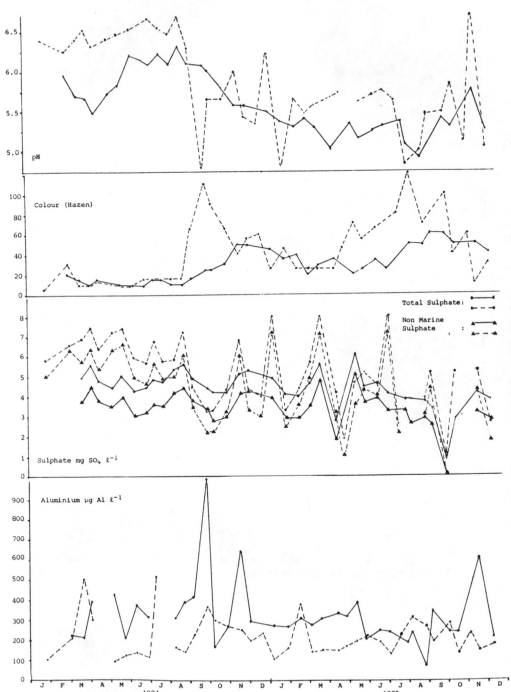

Fig. 4. Glendalough Lake Upper and its inflowing stream. Variation in the concentrations recorded in the lake (1 m depth) and in the inflowing stream (No. 1) for pH, colour, sulphate and aluminium in 1984 and 1985. Lake (1 m dpeth) ——— ; Inflowing Stream ---

55

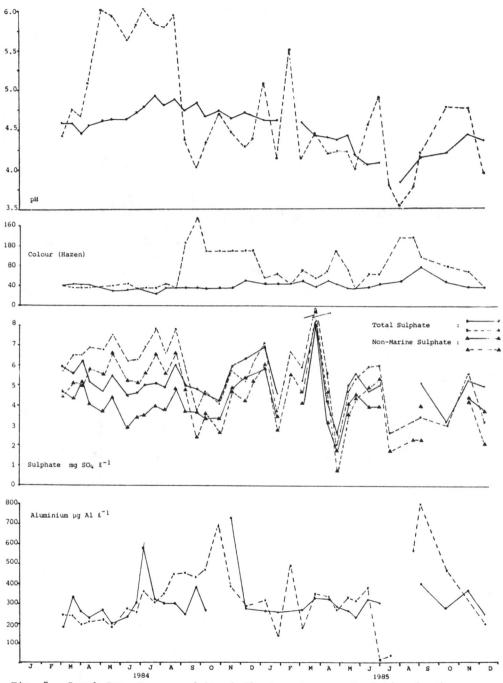

Fig. 5. Lough Bray Lower and its inflowing stream. Variation in the
concentrations recorded in the lake (1 m depth) and in the
inflowing stream for pH, colour, sulphate and aluminium in 1984
and 1985.
Lake (1 m depth ——— ; Inflowing stream ----

TABLE 4. The annual mean concentration for metals (total) (at 1 m depth) in Loughs Maumwee, Nahasleam, Bray Lower and Glendalough Lake Upper for 1984 and 1985.

Parameter / Lake	Year	Iron mg ℓ⁻¹ Fe	Sodium mg ℓ⁻¹ Na	Lead mg ℓ⁻¹ Pb	Potassium mg ℓ⁻¹ K	Manganese mg ℓ⁻¹ Mn	Cadmium mg ℓ⁻¹ Cd	Aluminium mg ℓ⁻¹ Al
Lough Maumwee	1984	0.062	11.46	0.007	0.443	0.011	0.0003	0.158
	1985	0.095	5.76	0.004	0.313	0.013	0.0004	0.091
Lough Nahasleam	1984	0.097	13.56	0.005	0.530	0.015	0.0004	0.136
	1985	0.168	7.70	0.004	0.355	0.016	0.0003	0.073
Glendalough Lake Upper	1984	0.146	4.34	0.014	0.31	0.063	0.0006	0.450
	1985	0.154	3.57	0.008	0.28	0.046	0.0006	0.254
Lough Bray Lower	1984	0.113	4.34	0.007	0.40	0.049	0.0003	0.419
	1985	0.131	4.29	0.004	0.38	0.041	0.0004	0.289

in Loughs Maumwee, Nahasleam and Glendalough Lake Upper ranged from 5.7 to 6.65 in 1984, while the minimum values were generally above 5.5.   In 1985, a decrease in the mean pH values of the above lakes, ranging from 0.25 to 0.6 of a pH unit, was noted.   The mean pH values in Lough Bray Lower were significantly lower than those in the other lakes and values of less than pH 4.0 were frequently recorded.

All the waters were markedly coloured on occasion with the maximum values occurring during the winter months.   Total phosphorus and orthophosphate concentrations were relatively low although the levels in Lough Bray Lower were somewhat higher than those in the other lakes. Ammonia, nitrite, silica and alkalinity concentrations were also low at all sampling points.   Nitrate concentrations were appreciably higher in the east coast lakes than in those of the west coast but in this case also the levels recorded were relatively low.

The most important anion associated with acidification of freshwaters is sulphate.   The concentrations of this compound were similar in the four lakes in 1984;   lower values were recorded in the following year, particularly in the west coast lakes.   However, the concentrations of sulphate of non-marine origin showed marked differences between the east and west coasts, the concentrations estimated in the former being over twice those in the latter.   The mean concentrations of "non-marine" sulphate recorded in the west coast lakes ($1.55 - 2.07$ mg SO_4 ℓ^{-1}, $32 - 43$ μ eq ℓ^{-1}) are at the lower end of the range of $30 - 60$ μ eq ℓ^{-1} estimated as the background levels for the Canadian Shield Lakes (National Research Council of Canada, 1981).

Detailed historical data for the physicochemical characteristics of the lakes studied are not available.   However, some short studies were carried out over thirty years ago notably by Webb (1947) on Loughs Bray and Glendalough Lake Upper;   Gorham (1957) performed a limited examination of the chemical composition of a number of poorly buffered west of Ireland lakes.   Webb reported pH values of 4.65 and 4.9 in the outlet from Lough Bray Lower and 6.25 in Glendalough Lake Upper.   He attributed the low pH values in some waters in the east of Ireland to the greater acidity of the peat in that area.   Gorham calculated the mean sulphate and pH values in a number of west of Ireland lakes as 6.7 and 5.2, respectively and reported values of 6.5 and 4.6 in Lough Shindilla, a lake adjoining Loughs Maumwee and Nahasleam.   Using the mean chloride concentration calculated by Gorham for the west of Ireland lakes a value of 3.2 mg SO_4 ℓ^{-1} may be estimated for "non-marine" sulphate in these waters at the time of Gorhams' measurements.   These values, when compared with those recorded in the present survey, suggest that there has been no significant increase of "non-marine" sulphate since the 1950s in Loughs Maumwee and Nahasleam.

The annual mean concentrations of the metals (total) recorded in Loughs Maumwee, Nahasleam, Bray Lower and Glendalough Lake Upper are given in Table 4.   The annual mean concentrations of aluminium in the west of Ireland lakes was less than 0.2 mg ℓ^{-1} Al whereas those in the east coast lakes ranged from 0.25 to 0.45 mg ℓ^{-1} Al.   Sodium concentrations were higher in the west coast lakes reflecting the greater marine influence in this area;   in contrast, manganese concentrations were highest on the east coast.   Concentrations of iron, lead, potassium and cadmium were generally low.

Biological Characteristics

The biological characteristics of one lake in each area, Lough Maumwee in the west and Glendalough Lake Upper in the east were examined. Phytoplankton.   Details of the phytoplankton populations encountered in

TABLE 5 : Comparative data on the phytoplankton populations of Maumwee
Lough and Glendalough Lake Upper for 1984 and 1985

	MAUMWEE LOUGH	GLENDALOUGH LAKE UPPER
Total Number of Taxa Identified[1]	231	108
Taxa occurring more than once[1] (Percentage of total in parentheses)	146 (63)	53 (49)
Mean number of Taxa per sample[1] 1984	40	11
1985	41	11
Number of CHLOROPHYCEAE (Excl. Desmidaceae)[1]	24	20
Number of DESMIDACEAE (Desmids)[1]	152	55
Number of BACILLARIOPHYTA (Diatoms)[1]	24	20
Number of CRYPTOPHYCEAE[1]	1	1
Number of CRYSOPHYCEAE[1]	8	5
Number of DINOPHYCEAE[1]	8	6
Number of CYANOPHYCEAE	14	5
Total number of Taxa identified in Enumerated Samples	98	46
Taxa occurring more than once in Enumerated Samples (Percentage of total enumerated in parenthesis)	52 (53)	23 (50)
Mean number of Taxa per enumerated sample 1984	11	6
1985	12	5
Maximum Cell Biomass $mm^3 \ m^{-3}$	170	76
1984 Mean " " " "	43	21
Minimum " " " "	13	5
Maximum Cell Biomass $mm^3 \ m^{-3}$	114	70
1985 Mean " " " "	30	17
Minimum " " " "	6	1

[1]
Numbers refer to the combined number of taxa recorded in both net and
Lugols preserved sample.

TABLE 6 : Abundance and number of taxa recorded for the principal
macroinvertebrate groups, on each sampling date in 1984 and
1985, in Maumwee Lough and Glendalough Lake Upper

GROUP		MAUMWEE LOUGH				GLENDALOUGH LAKE UPPER			
		1984		1985		1984		1985	
		11/7	5/12	16/5	7/11	9/7	17/12	14/5	5/11
PLECOPTERA	Number of Individuals	0	7	9	11	1	8	64	39
	% of Total Population	0	<1	<1	<1	<1	38	9	17
	Number of Taxa	0	2	1	4	1	1	7	5
EPHEMEROPTERA	Number of Individuals	21	2579	2715	1635	5	1	97	0
	% of Total Population	3	79	80	39	4	5	13	0
	Number of Taxa	2	10	8	8	1	1	3	0
CAENIS MOESTA	% of Total Population	2	60	66	33	0	0	0	0
TRICHOPTERA	Number of Individuals	36	43	70	43	12	2	38	71
	% of Total Population	5	1	2	1	10	10	5	31
	Number of Taxa	6	6	10	7	3	2	9	6
MOLLUSCA & CRUSTACEA	Number of Individuals	15	24	170	226	0	0	0	0
	% of Total Population	2	<1	5	5	-	-	-	-
	Number of Taxa	2	4	4	3	-	-	-	-
OLIGOCHAETA	Number of Individuals	134	64	69	871	54	6	219	56
	% of Total Population	18	2	2	20	46	28	30	25
	Number of Taxa	3	4	6	4	2	2	3	3
CHIRONOMIDAE	Number of Individuals	517	329	198	1430	33	0	234	53
	% of Total Population	70	10	6	34	28	-	32	23
	Number of Taxa	7	6	7	4	3	-	3	4
	Total Number of Individuals	739	3279	3398	4242	117	21	722	227
	Number of Taxa	20	32	36	30	10	6	25	18

Maumwee Lough and Glendalough Lake Upper are outlined in Table 5. A much more diverse phytoplankton standing crop was noted in Maumwee Lough than in Glendalough Lake Upper. In terms of the number of taxa recorded in the former lake the Chlorophyceae, principally Desmidaceae, were the principal forms; however the Dinophyceae, chiefly Peridinium inconspicuum and P. willei, were the main contributors to the algal biomass. Unidentified Cryptophyceae, diverse species of the Chlorophyceae and the Bacillariophyta, particularly Tabellaria spp., were less prominent in the overall biomass. The Crysophyceae, principally Dinobryon bavaricum, were prominent for prolonged periods.

In Glendalough Lake Upper the Chlorophyceae, particularly the Desmidaceae, were again most prominent in terms of numbers of taxa, but the Bacillariophyta, with Tabellaria spp. prominent on most occasions, and the Dinophyceae, with P. inconspicuum the principal form, were the main contributors to the algal biomass. Unidentified Cryptomonad spp. were also present in significant quantities.

There have been no previous studies on the planktonic algae in either lake. West and West (1904-1912) and West et al (1923) described studies they carried out on a number of west of Ireland lakes where they found very large and diverse desmid populations similar to those recorded in Lough Maumwee in the present survey. This suggests that little has changed in respect of the planktonic algae of that lake since the early part of the century.

The very low phytoplakton biomasses recorded in Maumwee Lough and Glendalough Lake Upper (maxima of 170 and 76 mm^3 m^{-3}, respectively) are well below the upper limit of 1000 mm^3 m^{-3} suggested for ultra-oligotrophic lakes by Vollenweider (1971). They are, however, of the same order as those described by Willen (1969) and Schindler and Nightswander (1970) for unacidified oligotrophic lakes.

Macrophytes. Macrophytes were poorly developed in both lakes. Lobelia dortmanna L. and to a lesser extent Potamogeton natans L. were the most prominent forms in the sub-littoral zones of Maumwee Lough. Phragmites communis Trin. and Juncus spp were present in sheltered bays. The very restricted littoral and sub-littoral zones of Glendalough Lake Upper may hinder the development of significant macrophyte stands. The principal plant present was Phragmites natans L.

Zooplankton. The diversity and abundance of Zooplankton recorded in Lough Maumwee were greater than those in Glendalough Lake Upper. In the collections from the former lake, 24 species of Rotifera, 20 species of Cladocera and 12 species of Copepoda were recorded compared, respectively, to 16, 10 and 3 in those from the latter lake. The principal Rotifera in Lough Maumwee were Conochilus unicornis, Kellicottia longispina, Keratella cochlearis and Polyarthra spp. The Cladocera, principally the Bosmina coregoni complex were frequently dominant in Maumwee Lough as was the Cyclopoid copepod Cyclops viridis. In Glendalough Lake Upper Keratella quadrata was the most prominent Rotifera with lesser numbers of Kellicottia longispina. The Cladocerans, B. coregoni complex, Holopedium gibberum and Diaphanosoma brachyurum were frequently the dominant forms in the lake. Immature Cyclopoid copepods were also present in significant numbers on occasion.

Macroinvertebrates. A comparison of the macroinvertebrate collections taken from Glendalough Lake Upper and Lough Maumwee is given in Table 6. Significant differences between the faunas of the lakes are apparent. The numbers of organisms taken in Maumwee Lough were on average ten times greater than in Glendalough Lake Upper, while the total number of taxa recorded was also appreciably greater in the former (77) compared to the

latter (48) lake.   The numerical dominance (<u>Caenis moesta</u> in particular)
and high diversity of the Ephemeroptera in Maumwee Lough, contrasts with
their relative paucity in Glendalough where <u>Leptophlebia</u> spp. was the
commonest genus.   Small numbers of Mollusca and Crustacea were recorded
in Maumwee Lough but these groups were not recorded in Glendalough.   The
greater abundance of the Plecoptera in the latter lake is in contrast
with the position for the foregoing groups.   Substantial populations of
Oligochaeta and Chironomeda were recorded in both lakes.

   <u>Fish</u>.   Brown trout (<u>Salmo trutta</u>), salmon fry (<u>S. salar</u>), minnow
(<u>Phoxinus phoxinus</u>) and eels (<u>Anguilla anguilla</u>) were captured in the
Maumwee Lough system while brown trout and minnow only were recorded in
the Glendalough Lake Upper system.   The examination of the brown trout
catch suggested that the population of this species in Lough Maumwee was
larger and faster growing than that in Glendalough Lake Upper.   However,
the measured growth rates from both lakes were relatively poor and
consistent with that recorded for other Irish lakes of similar trophic
status and geological characteristics.

   Trout, eels, minnow and salmon fry were captured in the Maumwee
feeder stream;   trout and minnow only were recorded in the principal
inflowing stream to Glendalough Lake Upper, and no fish were captured in
a second stream examined.   While trout numbers and growth rates were
higher in the inflowing stream to Lough Maumwee a larger standing crop
of trout was found in the inflowing stream to Glendalough Lake Upper due
to the presence of 0^+, 1^+, 2^+ and 3^+ trout in the latter.   The salmon
stocks in the Maumwee stream were all 0^+ fish.   Their mean length, weight
and numbers per m^2 were less than those for the 0^+ trout.

DISCUSSION

   The levels of acidity measured in the precipitation at sampling
locations on the west coast of Ireland were consistent with those for
relatively unpolluted areas.   The increased acidity noted in this area
during 1985, largely associated with rain bearing winds from an easterly
direction, indicated that under such climatic conditions, which are
normally infrequent as the prevailing wind is westerly, this area may
receive some artificially acidified rainfall.   Significant acidity was
measured on occasions in the precipitation at the east coast sampling
locations, particularly during 1985, and was also associated with
easterly winds.

   The pH values recorded in the two west of Ireland lakes, viz. Loughs
Maumwee and Nahasleam, were within the natural range for low alkalinity
waters;   it would appear from earlier measurements made on nearby lakes,
that there have not been any significant changes in the water chemistry
of the two study lakes over the past thirty years.   In contrast, the
minimum pH values recorded in Glendalough Lake Upper system in the east
of the country, particularly in 1985, were as low as those which have
been reported to cause damage to lakes ecosystems.   In this lake highly
significant correlations were found between the pH of the lake water and
both precipitation pH and lake water colour suggesting that both the runoff from the highly acidic peat and acidity in precipitation are
responsible for fluctuations in the acidity of the lake.   The "excess"
sulphate in the precipitation was significantly correlated with the "non-
marine" sulphate in the lake water further emphasising the influence of
the precipitation on the water chemistry of Glendalough Lake Upper.   In
Lough Maumwee the only significant correlation found was between lake
water pH and lake water colour which suggests that the acidity of the
lake water is principally determined by the humic acids leached into the

lake from the surrounding catchment.

While individual concentrations of aluminium (total) were high in Loughs Maumwee and Nahasleam the annual mean concentrations were less than 0.2 mg ℓ^{-1} Al, a concentration regarded as toxic to Brook trout. Concentrations in Glendalough Lake Upper were higher than 0.2 mg ℓ^{-1} Al; however, in the absence of information on the particular types of aluminium present further comment on the possible toxicity of the levels encountered to aquatic organisms is not possible. In the circumstances it is reasonable to assume that the greater part of the aluminium present was bound with organic ligands.

Lough Bray Lower is a dystrophic lake and the relatively marked acidity of its waters is due pricipally to run-off from the strongly acid peat in its catchment. In respect of its greater natural acidity it differs from the other lakes and, as such, is not suitable to be considered in the context of a baseline study of the chemical and biological characteristics of poorly buffered systems.

Significant differences between the biological characteristics of Lough Maumwee and Glendalough Lake Upper were noted. An appreciably greater diversity of phytoplankton and zooplankton populations was recorded in the former lake while certain acid-sensitive indicator zooplankton species which were present in Lough Maumwee, were not recorded in Glendalough Lake Upper. These differences suggest that the ecosystem of the west coast lake is not subject to acid stress whereas such an effect is present in the east coast lake. Further evidence of this difference between the two lakes is derived from the dominance of the sub-littoral macroinvertebrate communities of Lough Maumwee by acid sensitive forms such as Caenis moesta and other Ephemeroptera, Mollusca, Crustacea and Hirudinea, all of which were absent from Glendalough Lake Upper. In the latter lake the most prominent macroinvertebrate forms encountered were those regarded as acid tolerant and noted for their occurrence in artificially acidified aquatic habitats e.g. Leptophlebia vespertina and some Plecoptera.

The presence of salmon and trout stocks in the Lough Maumwee system which are similar to those in other Irish lakes of low alkalinity, suggests further that this lake is not subject to acid stress. While trout stocks are also present in Glendalough Lake Upper the nature of these stocks suggest that they are subjected to the sub-lethal stress. In consideration of the pH minima recorded in this lake it is possible that this stress is associated with the relatively high level of acidity.

Overall the chemical and biological characteristics of the Lough Maumwee system indicate that its poorly buffered waters, which would be highly sensitive to artificially acidified precipitation, are not so affected at this stage. The faunal and floral taxa present in Lough Maumwee, and the diversity of the communities encountered, are those characteristic of oligotrophic lakes unaffected by artificial acidification. The physicochemical characteristics of the precipitation recorded at the west of Ireland sampling locations, while indicating occasional episodes of increased acidity associated with rain bearing easterly winds, are comparable to those recorded in remote areas of Scandinavia and Canada which are free of the influence of atmospheric pollution. In contrast the corresponding data recorded for Glendalough Lake Upper on the east coast suggest that the lake is subjected to a significant degree of artificial acidification. It is suggested that the biological and chemical characteristics of Lough Maumwee would provide a useful baseline for assessing the impact of artificially acidic precipitation on other areas of Ireland or of other European countries.

ACKNOWLEDGEMENTS

The author gratefully acknowledges the assistance of his colleagues.
D. Butler, B. Callan, P. Fawl, M. Flanagan, H. Horan, K. Horkan,
M. Jackson, L. Lennox, C. O'Donnell and W. Wall.

REFERENCES

An Foras Forbartha, 1984.   Digest of Environmental Statistics, 1984.

Bailey, M. L. and Dowding P., 1985.   Distant and Local Pollutant
    Effects on Rain Chemistry and Leaf Yeasts.   An Foras
    Forbartha.

Bowman, J. J., 1986.   Precipitation Characteristics and the
    Chemistry and Biology of Poorly Buffered Irish Lakes :
    A Western European Baseline for "Acid Rain" Impact
    Assessment.   An Foras Forbartha.

Fisher, B. E. A., 1982.   Deposition of Sulphur and the Acidity of
    Precipitation over Ireland.   Atmos. Environ. 16. 2725 - 2734.

Gorham, E., 1957.   The Chemical Composition of Some Western Irish
    Fresh Waters.   Proc. R.I.A. Vol. 58, B.   237 - 243.

National Research Council Canada, 1981.   Acidification in the
    Canadian Aquatic Environment: Scientific Criteria for
    Assessing the Effects of Acidic Deposition on Aquatic
    Ecosystems.   367 pp.   NRCC/CNRC Canada.

Schindler, D. W. and Nighswander, J. E., 1970.   Nutrient Supply and
    Primary Production in Clear Lake, Eastern Ontario.
    J. Fish. Res. Bd. Canada 27 : 2009 - 2036.

Vollenweider, R. A., 1971.   Scientific Fundamentals of the
    Eutrophication of Lakes and Flowing Waters with particular
    Reference to Nitrogen and Phosphorus as Factors in
    Eutrophication.   Paris, O.E.C.D.

Webb, D. A., 1947.   Notes on the Acidity, Chloride Content, and
    Other Chemical Features of Some Irish Freshwaters.   Scient.
    Proc. Roy. Dublin Soc. 24. 24   215 - 228.

West, W. and West, G. S., 1904 - 1912.   The British Desmidiaceae
    Vols. I - IV.   London Ray. Society.

West, W. and West, G. S. and Carter, N., 1923.   The British
    Desmidiaceae. Vol. V.   London Ray. Society.

Willen, T., 1969.   Phytoplankton from Swedish Lakes II.   Assjon
    1961 - 1962.

ACIDIFICATION PROBLEMS OF FRESHWATERS: TROPHIC RELATIONSHIPS

D.W. Sutcliffe, T.R. Carrick, A.C. Charmier, T. Gledhill, J.G. Jones,
A.F.H. Marker and L.G. Willoughby
Freshwater Biological Association, The Ferry House, Ambleside,
Cumbria LA22 OLP

Summary

Some major features of the chemical composition of "soft" (mean pH c. 7.0 to 5.5) and "acid" (mean pH < 5.5) waters are briefly summarized in a desk study, including ionic requirements and regulation by animals, numbers of macrophyte plant species and invertebrates, and feeding guilds, in acid and soft waters. The desk study is followed by brief descriptions of the flora of streams in the River Duddon, microbial decomposition of detritus, and field and laboratory observations on feeding and acid-tolerance of some representative herbivorous benthic invertebrates.

1. GENERAL INTRODUCTION

The work described in this report was carried out under Contract ENV. 865.UK(H) for the Commission of the European Communities, for an 18-month period commencing 1 November 1984. The main objective was to examine the foods and physiological requirements of selected herbivorous benthic invertebrates living in "acid" waters (pH below 5.5) and in circumneutral waters (pH 6.5-7.0), chiefly in streams of the River Duddon catchment in the English Lake District. Survival with and without food was also examined in laboratory experiments. Detailed in-depth studies were not attempted. The primary concern was to examine a number of aspects of invertebrate feeding and physiology, related to chemical composition of the water, and to identify those aspects that are most important in relation to the acidification of streams (and to a certain extent lakes). This would provide a basis for deciding which aspects could and should be examined in greater detail in future work.

2. FIELD AND LABORATORY STUDIES

In upland acid streams of the River Duddon, the epilithon on stones consisted mainly of four filamentous algae; relative abundances of each varied between streams. These were thick-sheathed, blue-green Cyanophyceae, Homeothrix caespitosa and Scytonema crispum, and green filamentous Chlorophyceae, Hormidium subtile and Zygogonium sp. All were present in the gut contents of acid-tolerant herbivorous invertebrates but, in the laboratory, Hormidium appears to be the most acceptable as a food, sustaining fast growth by larvae of an acid-tolerant stonefly, Amphinemura sulcicollis, and a mayfly, Siphlonurus lacustris. Both animals fed extensively on Hormidium in the acid streams, and so did Simulium.

Algae were quantified by measuring chlorophyll-a. The largest amounts were found in lowland soft streams without a canopy of trees.

Here the commonest algae were Rhodochorton (Rhodophyceae) and diatoms. The latter predominated in soft streams overshadowed by trees but the diatoms were sparse; amounts of chlorphyll-a were lower than in upland acid streams without trees. Here, filamentous algae were dominant; the only common diatom was Tabellaria, eaten by herbivores but sparse in numbers.

A liverwort, Nardia compressa, predominates in acid streams whereas a moss, Hygrohypnum luridum, predominates in soft streams. Epiphytic diatoms, Eunotia and Peronia, are numerous in the leaf axils of Nardia, where they are relatively inaccessible to grazing animals. The diatom Cocconeis is abundantly epiphytic on the flat leaves of Hygrohypnum, where it is accessible to grazers; Cocconeis and another large diatom, Gomphonema, feature prominently in the diets of mayfly larvae in soft streams. The absence of these diatoms from acid streams, also Rhodochorton which is extensively grazed upon in soft streams, may account for the absence of some algivorous mayflies, e.g. Baetis and Ephemerella. Nevertheless Baetis rhodani does occur in acid streams of the English Peak District, where concentrations of major ions, especially calcium, are particularly high.

The other main food source for herbivores in the River Duddon is decomposing organic matter (detritus). Where trees are scarce, as in the upper Duddon, allochthonous sources of detritus are mostly blown-in dead leaves of grasses, chiefly Nardus stricta, and bracken, Pteridium aquilinum. In field trials with leaves of alder, oak and Nardus, decomposition rates were measured by loss in weight. Oak and Nardus decomposed twice as fast and alder six times as fast in soft streams, compared with acid streams. In the latter, leaves were colonized by fewer hyphomycete fungi and the bacterial flora was distinguished by a lack of Flavobacterium, Cytophaga and Flexibacter. These bacteria declined in numbers when colonized leaves were transferred from a soft to an acid stream, demonstrating the selective nature of acid water.

In acid streams the stonefly larva Amphinemura sulcicollis feeds on Nardus and Sphagnum leaves, and on the algae Hormidium, Zygogonium and Tabellaria. In the laboratory, Amphinemura (and the freshwater shrimp Gammarus pulex) grow very slowly when fed on Nardus, presented as small fragments of leaves or as "fine" detritus. When passed through a sieve to exclude particles more than 150 μm in diameter, fine detritus collected from an acid stream did not support survival or growth of the mayfly larva Baetis muticus, kept in soft water (pH 7.1). Survival was improved when B. muticus was given fine detritus collected from a lowland soft stream, where the animal feeds mainly on the detritus derived from leaves of deciduous trees.

At 10°C, both B. muticus and B. rhodani survived (unfed) for only 2 to 8 days in acid streamwaters at pH 4.9 to 5.2. Survival of B. rhodani was better (100% survival after 8 days but 100% mortality after 11 days) in an acid tarnwater at pH 4.8; this contained unusually high concentrations of major ions, including calcium. Experimental survival was similarly improved by adding calcium (0.5 and 1.0 mE l^{-1}) to acid streamwaters of low ionic strengths.

3. RECOMMENDATIONS FOR FUTURE RESEARCH

The following topics would be worth pursuing in depth. They are listed here for general consideration; it is not implied that all could or should be carried out specifically in the River Duddon or, indeed, at the FBA. (Each recommendation is cross referenced to the relevant section in the full Project Report submitted to the Commission).

1. Microbial "conditioning" and decomposition of leaf litter. The roles of hyphomycete fungi and bacteria in the breakdown of leaf litter in streams are still not known. Mutual interactions between fungi and bacteria may be important, each organism providing specific enzymes necessary for decomposing cell walls. Both fungi and bacteria may be sensitive to low pH (< 5.5). To what extent do their activities increase the nutritional values of organic detritus as well as improving its palatability to invertebrates? (Sections 3.3, 3.4, 3.5, 3.8, 4.2, 4.3).

2. Digestive enzymes in the guts of herbivorous invertebrates, especially cellulases. (Section 3.9).

3. The formation, composition and structure of epilithon and its importance in the feeding of herbivorous invertebrates, compared with epiphyton. (Sections 2.5, 3.2, 5.3, 5.4, 5.5).

4. Community structure in terms of assemblages of species, trophic levels, biomass, and the role of predators. (Sections 2.1, 2.2, 2.4, 2.5).

5. Ionic regulation in acid water, accumulation of toxic and non-toxic metals, moulting, drinking and excretion, and the importance of foods in supplying essential ions. (Sections 2.2, 2.3, 3.7).

6. Moulting, metabolism and growth; differing metabolic requirements of acid tolerant and intolerant species. (Sections 2.3, 3.3, 3.6, 3.8).

PHYSIOLOGICAL STUDY ON THE RECOVERY OF RAINBOW TROUT

(SALMO GAIRDNERI RICHARDSON) FROM ACID AND AL STRESS

H.E. WITTERS [*] °, J.H.D. VANGENECHTEN [*], S. [*] VAN PUYMBROECK [*]
and O.L.J. VANDERBORGHT °
* Belgian Nuclear Center, Department of Biology,
  Boeretang 200, B-2400 Mol, Belgium
° University of Antwerp, Department of Biology,
  Universiteitsplein 1, B-2610 Wilrijk, Belgium

Summary

It is known that the ionoregulatory system of fishes is seriously
affected in acid soft waters with or without elevated
Al-concentrations. High branchial ion losses are supposed to result
from Ca-modulated increases in the permeability of the branchial
epithelium.
The results of preliminary experiments on the recovery of rainbow
trout after exposure to acid water (pH 5.0) or to acid water with
Al (pH 5.0 + 100 µg Al/l) are presented. Recovery was studied at pH
6.9 with low (1.0 mg/l) and high (19.0 mg/l) Ca-levels.
Compared to acid stress alone, fish were more affected by the
combined stress of acidity and Al (90% mortality, severe Na- and
Cl- whole body loss). However, indications were obtained that
the depleted plasma Na-content increased faster to normal values
after combined acid + Al stress. Indeed acid exposed fish were not
able to restore the normal plasma ionconcentrations after 7 days of
recovery. High Ca-levels in the water did not favor recovery in these
experiments.

1. INTRODUCTION

Recent research on the physiology of environmental acid- and
Al-stress in aquatic organisms has mostly focussed on ionregulation,
acid-base control and respiratory processes in fish (1,2,3,4). These
studies have indicated that the ionregulatory processes are strongly
affected by acid conditions in soft waters. The impaired ionbalance in
acid- and Al-exposed fish is mostly due to an elevated branchial ion loss
of Na, Cl, K and Ca (2,3,4). The study of McWilliams and Potts (5)
provides good evidence that the permeability of the branchial epithelium
is altered in the presence of high hydrogen ion levels. It is proposed
that the paracellular pathway in tight epithelia (e.g. fish gill) is
opened at low external pH as was already observed in vitro in frog skin
(2,6). This opening of paracellular channels could be explained by the
displacement of calcium ions from the binding sites at the junctions
(2,6).
The objective of the present study was to investigate if and how the
disturbance of the ion balance in acid- and Al-stressed fish can be
overcome when water pH is increased again to circumneutral values. The
role of Ca, as a modulator of the gill permeability, is tested during the
recovery process.

2. MATERIALS AND METHODS

The experiments were carried out with rainbow trout, <u>Salmo gairdneri</u> Richardson (1.5 year old, 180-320 g). These fish were obtained from a hatchery in France. Prior to the experiments the fishes were acclimated for 8 to 12 days at 8°C in a closed water recirculation system (125 l total volume; 1000 l/hr). The acclimation water (table 1), which was of the same ionic composition and pH (pH 6.9) as the control water in the experiments, was daily renewed to minimize accumulation of excretory products. For this reason too, fishes were starved during the acclimation period and during the course of the experiments.

The experimental solutions were prepared by adding some salts (NaCl, KCl, CaCl$_2$, NaHCO$_3$ and MgSO$_4$.7H$_2$O) to demineralized water (table 1). Higher Ca-levels were made by addition of Ca(NO$_3$)$_2$.4H$_2$O. To acidify the solutions a mixture of 2 volumes H$_2$SO$_4$ (0.1 N) and 1 volume HNO$_3$ (0.1 N) was added. The pH was kept constant at pH 5.0 $\pm$ 0.1 by an automatic pH-stat device. Al was added to the acid solutions as AlCl$_3$ to reach a final concentration of 100 µg Al/l. As Al disappears readily from the water, freshly prepared water (100 µg Al/l) was constantly added (25 l/hr) to the recirculation system.

After acclimation, the fish were exposed for 6 days to 3 different experimental stress conditions as indicated in figure 1. The stress period was followed by a recovery period of 7 days (fig.1). The water (control, acid or acid + Al) was changed at the end of the stress period by rinsing the test chambers several times with control water of either low Ca content (1.0 mg/l) or high Ca content (19 mg/l) (table 1).

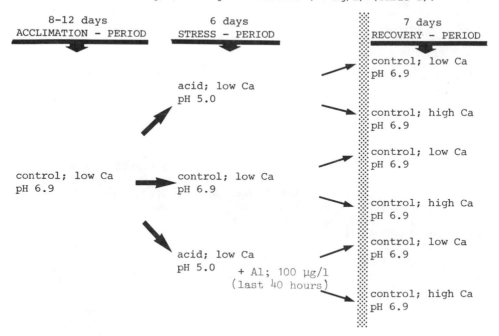

Figure 1. Scheme of the experimental protocol.

At different time intervals during the 6 day stress period and the subsequent 7 day recovery period, whole body net ionfluxes were measured and blood samples were taken. Arterial blood was sampled from fishes

Table 1. Ionic composition (mg/l) and pH in the acclimation water and experimental solutions (mean ± 95% confidence limits).

	Ca	K	Mg	Na	Cl	HCO$_3$	pH
acclimation water	1.10	0.98	0.27	9.28	9.80	14.4	6.85
stress							
* acid	1.15 ± 0.10	1.05 ± 0.10	0.27 ± 0.001	9.38 ± 0.03	9.94 ± 0.29	4.4 ± 0.1	5.0 ± 0.1
* acid + Al	1.22 ± 0.03	1.03 ± 0.03	0.31 ± 0.006	9.44 ± 0.19	11.39 ± 0.14	4.3 ± 0.1	5.0 ± 0.1
recovery							
* low Ca-control	1.01 ± 0.03	0.71 ± 0.04	0.26 ± 0.003	8.93 ± 0.14	9.00 ± 0.12	17.1 ± 0.1	6.9 ± 0.1
* high Ca-control	18.85 ± 0.22	0.74 ± 0.05	0.27 ± 0.004	8.86 ± 0.15	7.02 ± 0.14	18.6 ± 0.1	7.0 ± 0.1

fitted with a catheter in the dorsal aorta. Surgery was carried out 24 hours before the first blood sample was drawn. The dorsal aorta cannulation technique was used to allow repeated sampling in unanaesthetised fish during the whole course of the experiment. The technique described by Soivio (7,8) was followed with minor modifications for the preparation as well as for the surgery.

Fish were kept in individual test chambers (3.5 to 4.0 l) during the experiment (4). The whole body ionfluxes were measured over a period of 2.5 hours during which the waterflow was stopped in the test chamber. Watersamples were taken at the beginning and at the end of the flux period (4). Blood samples (200-400 µl) were drawn just before the start of the whole body ionflux measurements.

The ionconcentrations in the watersamples were measured using a plasma emission spectrophotometer (Jarrel Ash, Atomcomp,Model 750) for the cations and a segmental flow analyser (Skalar,Model 5100) for the anions (4). The different species of Al in the water were determined using the method of Driscoll (9). The blood samples were centrifuged in a refrigerated (4°C) centrifuge (Heraeus,Minifuge GL). Plasma was diluted (1:100) to measure cation- and anionconcentrations (4).

The whole body ionfluxes were calculated using the equation:

$$\frac{(Cf-Ci) \times V}{W \times t} \ (\mu equiv/kg.hr)$$

Cf is the final and Ci the initial ionconcentraion (µequiv/l) in the fish chamber, V is the watervolume (l), W is the weight of the fish (kg) and t is the duration of the measurement period (2.5 hours). It is estimated that a difference of 100 µequiv/kg.hr in the Na and Cl whole body fluxes can be detected with reasonable accuracy.

The table indicates mean values ± 95% confidence limits. The values in the figures are means ± standard error on the mean. The student two-tailed 't'-test was used to test statistical differences. The chosen level of significance is P < 0.05

3. RESULTS

The figures (2,4) indicate that the ionfluxes are not disturbed in acid water (pH 5.O). During the recovery period at low and high Ca-levels, only minor differences are observed between the acid-exposed and the control-exposed fish. Since these differences are less than 150 µequiv/kg.hr in almost all cases, they are not regarded as meaningful (see materials and methods). The Na- and Cl-concentration in the plasma (fig.3 and 5) however, are significantly lower in acid-stressed fish compared to the control group. The presence of high ambient Ca-levels (19 mg/l) did not seem to have any stimulating effect on the restoration of the plasma ionconcentrations in acid-stressed fish.

The exposure to Al at pH 5.O caused a high mortality rate of the fish. The acid + Al group contained 41 trout, from which 35 fish died gradually during the 40 hours of acid + Al-exposure. Another two fish died within the first 3 hours of the recovery period. The measurement of Al indicated a total water concentration of about 100 µg Al/l of which 30 to 40 µg/l was retained on a 0.45 µm Millipore membrane filter. The extraction with methyl isobuthylketone (9) indicated that 100% of the "dissolved" fraction (non-retained on the filter) was labile monomeric aluminium. This Al-form is supposed to be the toxic fraction of Al to aquatic organisms (9). It is supposed that the filter-retained Al-fraction

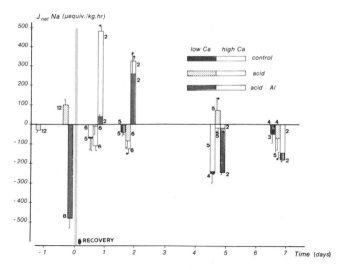

Figure 2. The whole body Na flux in rainbow trout during a period of
stress (pH 5.0 or pH 5.0 + 100 µg Al/l) and a subsequent
recovery period. The start of the recovery is indicated by the
bar at day 0. All values are means ± S.E.M. The numbers are the
number of experimental animals. Asterisks indicate mean values
significantly different (* P < 0.05) from corresponding control
values (same Ca-levels), tested by two tailed 't'-test.

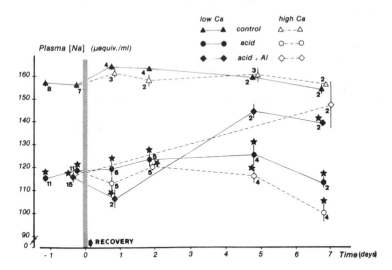

Figure 3. The plasma Na-concentration in rainbow trout during a period of
stress (pH 5.0 or pH 5.0 + 100 µg Al/l) and a subsequent
recovery period. The start of the recovery is indicated by the
bar at day 0. All values are means ± S.E.M. The numbers are the
number of experimental animals. Asterisks indicate mean values
significantly different (* P < 0.05) from corresponding control
values (same Ca-levels), tested by two tailed 't'-test.

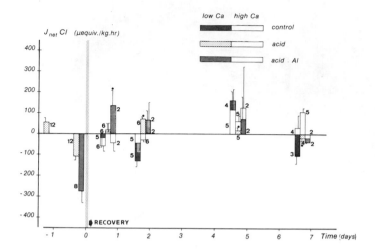

Figure 4. The whole body Cl flux in rainbow trout during a period of stress (pH 5.0 or pH 5.0 + 100 µg Al/l) and a subsequent recovery period. The start of the recovery is indicated by the bar at day 0. All values are means ± S.E.M. The numbers are the number of experimental animals. Asterisks indicate mean values significantly different (* P < 0.05) from corresponding control values (same Ca-levels), tested by two tailed 't'-test.

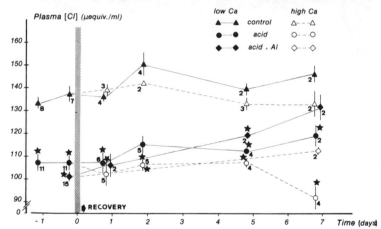

Figure 5. The plasma Cl-concentration in rainbow trout during a period of stress (pH 5.0 or pH 5.0 + 100 µg Al/l) and a subsequent recovery period. The start of the recovery is indicated by the bar at day 0. All values are means ± S.E.M. The numbers are the number of experimental animals. Asterisks indicate mean values significantly different (* P < 0.05) from corresponding control values (same Ca-levels), tested by two tailed 't'-test.

(30-40 μg Al/l) is under colloïdal form,e.g. strongly adsorbed to the mucus secreted by the animals in the acid water containing Al.

Figures 2 and 4 indicate that the exposure to Al at low pH caused a high loss of Na and Cl from the body which is significantly higher than observed in fish exposed to pH 5.0 alone. The exposure to control water of pH 6.9 during recovery induced a net uptake of Na in acid + Al-stressed fish after 1 day of recovery. This is more pronounced in high Ca water. At the 2nd day of recovery the net gain of Na is about the same at low and high Ca-levels. After 6 to 7 days the whole body Na flux in acid + Al-stressed fish is of about the same magnitude as the values in control fishes. A whole body Cl loss is noticed at acid + Al-exposure. At low Ca a net uptake of Cl is observed which is comparable to control values. Opposite to the remarkable effect of high Ca-levels on the Na-flux at the first day of recovery (fig.2), no such effect is observed on the Cl-whole body flux (fig.4).

The plasma Na- and Cl-concentrations (fig.3 and 5) are significantly reduced in acid + Al-exposed fish. After the 7 day recovery in control water with high Ca-levels the plasma Na- and Cl-concentrations are the same as in control fish. At low Ca-levels only the Cl-concentration in the plasma is restored.

4. DISCUSSION

4.1. Acid exposure

Fig. 3 and 5 indicate that plasma Na- and Cl-concentrations are significantly reduced after 6 to 7 days of acid exposure compared to values measured in control fish. This observation is in agreement with earlier acid exposure studies (2,3,4). The Na-concentration in the plasma of acid exposed fish is about 26% lower than control values. The plasma Na-concentration remained significantly lower than control levels during 7 days of recovery. This can be explained by the observations on the whole body Na flux (fig.2). Taking into account the variation of the whole body ion fluxes (100 μequiv/kg.hr) as a result of the method of the measurements, it is seen that in acid exposed fish whole body fluxes are comparable to those in control fishes. Because no net gain of Na is noticed during recovery, the plasma Na-concentrations should stay almost the same as is confirmed by our results (fig.3). There is apparently no ameliorating effect of Ca on either plasma Na concentration or the Na whole body flux during the recovery period.

The effect of low pH on Cl-balance seems to be less pronounced. The plasma Cl-concentration is decreased with 17% during acid stress. The plasma Cl-concentration was not reestablished during the recovery period. This can be explained by the whole body Cl-fluxes which are comparable to values in control fish and no net gain of Cl as a compensation for low plasma Cl-levels, is noticed. Ca did not seem to have any effect on the recovery of the Cl-balance in acid exposed fish.

4.2. Acid + Al exposure

Notwithstanding the fact that experimental conditions seemed to be severe (90% mortality in acid + Al exposed fish), some fish recovered. The plasma Na-concentrations are significantly reduced as a consequence of the high Na-loss at acid + Al-exposure. During the 7 day recovery at low Ca-levels, a gradual increase of the plasma Na-concentration is seen (fig.3). The plasma Na-concentration is also restored at high Ca-conditions so that no significant difference between acid + Al-exposed fish and control fish is observed anymore at the 7th day of recovery. It is clear from the whole body Na flux measurements (fig.2) that the restoration of the plasma Na-concentration can be explained by the significant net gain of Na on the first (high Ca) and second (low and high Ca) day of recovery. It is calculated that the net gain of Na at the first and second day of the recovery period (high Ca), is 7 times higher than necessary to restore the decreased plasma Na concentration. The impact of the net gain of Na on the other body compartments is not measured.

The Cl-balance was also disturbed by the 40 hours of acid + Al-exposure. The plasma Cl-concentration was 25% lower in acid + Al-exposed fish compared to control fish (fig.5). But after 7 days of exposure to control water (pH 6.9) the plasma Cl-concentration was not significant different from levels measured in control fish. The recovery of the fish, observed as a restoration of plasma Cl-levels, can be explained by the elevated net uptake of Cl at low Ca-levels. However no net gain of Cl is observed at high Ca-conditions so that it remains unexplained how plasma Cl-concentration was restored in these conditions. It is supposed that the high nitrate levels (calculated: $\pm$ 70 mg/l) at high Ca conditions could interfere with the Cl regulation. However no literature data were found mentioning such effects of nirate to fish.

5. CONCLUSION

We emphasize the preliminary aspect of this study due to the low numbers of fish in some experimental groups. Nevertheless it is observed that * trout are more affected by the combined stress of low pH (pH 5.0) and Al (100 µg/l) (90% mortality) compared to acid stress (pH 5.0) alone.
* fish exposed to acid water with Al recovered from stress conditions in opposite to acid exposed fish.
* the restoration of the plasma Na levels in acid + Al exposed fish was possible due to a net gain of Na during the first two days of recovery. It is calculated that the net gain of Na is several times higher than necessary to restore plasma ion levels.
* recovery conditions with high Ca levels in comparison with low Ca levels gave no indications of a possible role of Ca as a modulator of gill permeability.

REFERENCES

(1) ROSSELAND, B.O. (1980). Physiological responses to acid water in fish. 2. Effects of acid water on metabolism and gill ventilation in brown trout, Salmo trutta L., and brook trout, Salvelinus fontinalis M. In : Proc. Int. Conf. Ecol. Impact Acid Precip. Norway, SNSF Project, eds. D. Drabløs & A. Tollan, 348-349.
(2) McDONALD, D.G., WALKER, R.L. and WILKES, P.R.H. (1983). The interaction of environmental calcium and low pH on the physiology of the rainbow trout. II. Branchial ionoregulatory mechanisms. J. Exp. Biol. 102, 141-155.
(3) HÔBE, H., WOOD, C.M. and McMAHON, B.R. (1984). Mechanisms of acid-base and ionregulation in white suckers (Catostomus commersoni) in natural soft water. I. Acute exposure to low ambient pH. J. Comp. Physiol. B. 154, 34-46.
(4) WITTERS, H.E. (1986). Acute acid exposure of rainbow trout, Salmo gairdneri Richardson : effects of aluminium and calcium on ion balance and haematology. Aquatic Toxicology, in press.

(5) McWILLIAMS, P.G. and POTTS, W.T.W. (1978). The effects of pH and calcium concentrations on gill potentials in the brown trout, Salmo trutta. J. Comp. Physiol. 126, 277-286.

(6) GONZALEZ, E., KIRCHHAUSEN, T., LINARES, H. and WHITTENBURG, G. (1976). Observations on the action of urea and other substances in opening the paracellular pathway in amphibian skins. In : Comparative Physiology : Water, Ions and Fluid Mechanisms, eds. K. Schmidt-Nielson, L. Bolis & S.H.P. Madrell, 43-52.

(7) SOIVIO, A., WESTMAN, K. and NYHOLM, K. (1972). Improved method of dorsal aorta catheterization haematological effects followed for three weeks in rainbow trout (Salmo gairdneri). Finnish Fish Res. 1, 11-21.

(8) SOIVIO, A., NYHOLM, K. and WESTMAN, K. (1975). A technique for repeated sampling of the blood of individual resting fish. J. Exp. Biol. 62, 207-217.

(9) DRISCOLL, C.T. (1984). A procedure for the fractionation of aqueous aluminium in dilute acidic waters. Intern. J. Environ. Anal. Chem. 16, 267-283.

THE DEVELOPMENT OF THE ACID LAKE GRIBSØ IN DENMARK AFTER 1950

by

S. Wium-Andersen
Copenhagen University
Freshwater Biological Laboratory,
51 Helsingørsgade, DK-3400 Hillerød
Denmark.

Abstract
Lake Gribsø is a polyhumic acid lake situated in North Zealand,
Denmark. The pH of the tributaries is today, as in 1950, in the
range of 3.2 to 3.9. The mean acidity of the tributaries is today 1
meq/l. The pH of the epilimnion of the lake dropped from 5.2 in 1950
to 4.3 in 1985 Secchi disc depth decreased from summer values around
3 metres in 1950 to 1 metre in 1985. The acidification seems to have
accelerated during the last 35 years.
Both flora and fauna have changed during the last century.All species of Melosira disappeared between 1919 and 1926. Many other phytoplankton species have disappeared, and only a few new ones have
occurred. Today the phytoplankton is dominated by very few species
usually occurring in monocultures, e.g. Synura petersenii and S.
sphagnicola, Dictyosphaerium pulchella and a species of Chlamydomonas. Holopedium gibberum disappeared between 1868 and 1900. The
number of Bosmina and Daphnia were reduced from 1920 to 1940 and
they disappeared alltogether before 1985 together with Eudiaptomus.
The rotifers Conochilus, the water mites Piona sp., were very common
in 1940 but were not found in 1985. The bottom living animals Asellus, Pisidium, Herbopdella and Limnophilus were recorded from the
lake 30 years ago but not in 1985. There is still perch and pike in
the lake, but fry has not been seen.

Introduction
    The effect of the recent acidification of lakes and rivers has only
been the topic for rather few studies in Denmark. Rebsdorf (1983) reported a study on oligotrophic seepage lakes in Jutland and demonstrated a
tendency towards acidification. He believes that the acidification was
caused mainly by atmospheric depositions of acidifying substances.
    Older studies have been dealing with the acidification of the well
studied Danish lake, Gribsø, in North Zealand. The first more comprehensive description of Lake Gribsø, was made by Wesenberg-Lund (1940). Berg
et al. (1956) published a monograph about the lake "Studies on the humic
acid lake Gribsø", with studies of the fauna in 1941-1942 and some basic
hydrographic observations. Nygaard (1965) published further information
about the chemical composition of the lake water from 1949-1951.
    This paper reports a part of a study carried out on the lake from
1984 to 1986. This study has shown that the acidification has been
accelerating during the last 35 years.

Description of the area

Lake Gribsø is situated in North Zealand, Denmark. The size of the lake is 10.1 hectares, the volume 484 000m^3, the mean depth 4.8 metres, and the maximum depth is 11 metres. The lake is a secondary water table situated 50 metres above sea level. Primary ground water level is about 35 metres below the lake surface. The theoretical catchment area is 180 hectares, but about two thirds of the water seeps into the ground, down to the primary water tabel.

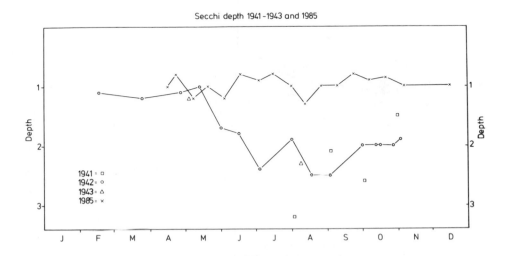

Fig. 1. Secchi disc depth in Lake Gribsø in 1941-43 and 1985. Values from 1941-43 from table 14 in Berg et al.(1956).

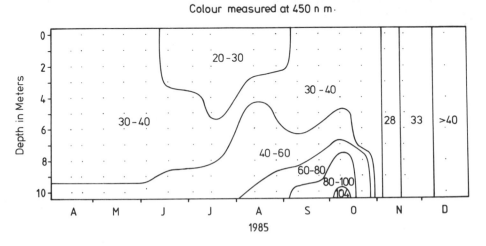

Fig. 2. Water colour measured as absorbance at 450 nm in Lake Gribsø. Every dot in this and the following figures indicate one measurement. 10 mm cuvette.

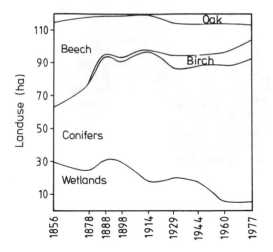

Fig. 3. Land use in the area surrounding Lake Gribsø from 1856 to 1977 based on forest maps.

Lake Gribsø receives water from the bogs in the vicinity through 6-7 small brooks, which carry water only in spring and autumn. The total inflow was in the range of 120 000 m^3 in 1985. The inflowing water is very acid, pH 3.2 - 3.9, acidity around 1 meq/l. The colour of the water is as strong tea due to humic substances. This has also resulted in a change in the Secchi disc depths in the lake (fig. 1). In 1940 a clearwater period was recorded in summer,this is not seen any more. However it can still be recognised if measurements are carried out on the colour of the water at 450nm. (fig. 2). There is no outlet from the lake. Changes in the water level are in the range of 0.6 metres.

Land use around the lake is shown in fig. 3. The lake is surrounded by a plantation. The catchment area is covered with 77% spruce, 22% deciduous trees and 5% bogs. No sewage is entering the lake. For further information see Berg et al.(1956).

Results and discussion

Lake Gribsø is stratified from early spring to late autumn due to the surrounding hills and forest. (fig 4). The gross primary production was 116 gC/m^2 in 1985, (fig 5), and the allochtonous input is in the same range. Due to the oxygen demand from degradation of these components the hypolimnion is depleted of oxygen from July until the autumn overturn. (fig 6). The oxygen depletion combined with the very distinct stratification results in a heavy build up of CO_2 in the hypolimnion. (fig 7).

Nutrient concentrations in the lake have increased since 1950, especially ammonium and phosphate. Primary production is limited to the upper 2 metres of the water column due to the very high content of humic substances in the lake, and occasionally by the very low CO_2 content.

The pH in the tributaries to the lake is shown in fig 8. In most of the tributarys is the pH below 4, often between 3.2 and 3.9. Unpublished measurements by Nygaard in 1949-50, carried out with the same type of equipment as used to day (Nygaard 1965), are also shown in the figure. The observations indicates, that the pH in the tributariess has been rather constant through the last 35 years. Today the acidity of the tribu-

taries is in the range of 1 meq/l. One third of the acidity is in strong
acids, and 2/3 on weak acids according to unpublished results by Pedersen
and Sindballe(Freshwater Biological Laboratory, Copenhagen University).

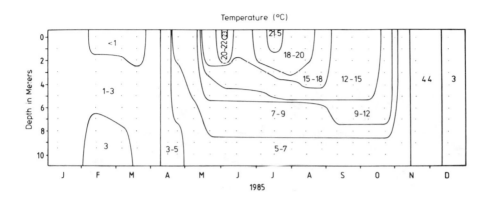

Fig. 4. Temperature measurements from Lake Gribsø, 1985.

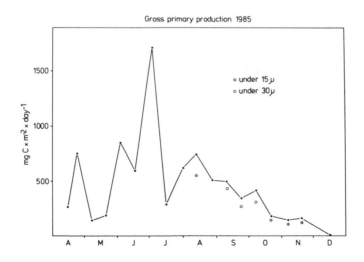

Fig. 5. Gross primary production. Production of the nannoplankton
         is indicated.

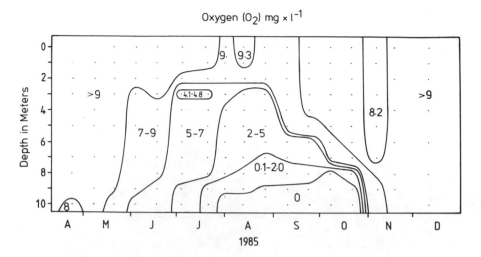

Fig. 6. Oxygen measurements from 1985 in mgO$_2$/liter.

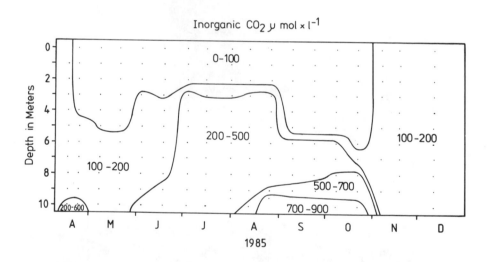

Fig. 7. Inorganic CO$_2$ in µmol/liter. Measurements carried out with infrared gas analysis.

The pH of the lake in 1949-50 and in 1985 is shown in figs 9 and 10. In the epilimnion the pH is about 1 unit lower today than 35 years ago. (fig 11). The very high pH recorded in the bottom water is due to a reduction of sulfate to HS$^-$. Just before the overturn the water has a very strong smell of hydrogen sulfide.

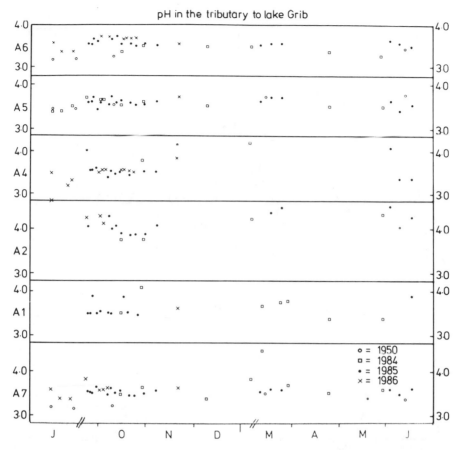

Fig. 8. pH in the tributaries to Lake Gribsø. The tributaries are named
A-1 to A-7 according to Fig. 2 in Berg et al. (1956).

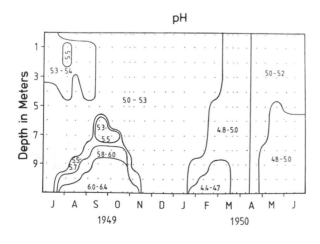

Fig. 9. pH in Lake Grib-
sø in 1949-50.
This figure is
drawn after values
from table 28 in
Nygaard(1965).

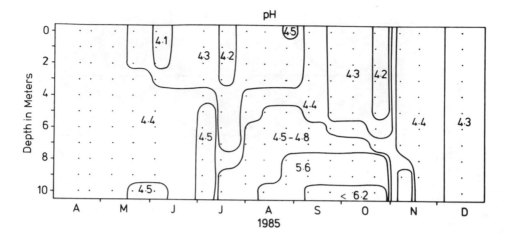

Fig.10. pH in Lake Gribsø 1985.

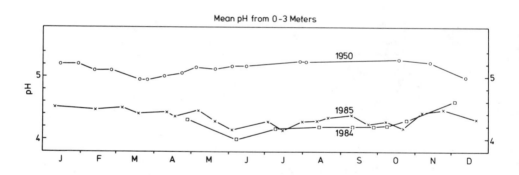

Fig.11. Mean pH from 0-3 metres in Lake Gribsø in 1950-1985. Values from
1950 are from Table 28 in Nygaard (1965).

The first record of an animal disappeared from the lake was Holopedium gibberum. This species was collected in the lake by Müller(1868),
and in 1902 Wesenberg-Lund(1904) searched for it in vain. Wesenberg-Lund
(1940) discussed the acidification in relation to the decrease in the
numbers of Daphnia longispina and Bosmina longirostis in the lake since
1920 and predicted the disapperance of these species.
    Bosmina, Daphnia, Eudiaptomus and the water mite Piona were all re-
corded by Berg in his investigations in 1941-1942, and Conochilus was recorded in 1949 by Nygaard (see Berg et al.1956). The five mentioned species were not recorded in 1985 when plankton samples were collected every
fortnight with a 80 μm net. The zooplankton is now dominated by Cerio-
daphnia and rotifers, e.g. Synchaeta sp.

There is still pike and perch in the lake. They are extremly slow
growing and they don't seem to breed since fry has not been seen in the
lake for years. The top predator of the system today is the phantom midge
Chaoborus flavicans. These changes in the zooplankton and fish community
are in accordance with that recorded from Swedish lakes (Nymann et
al.1985).

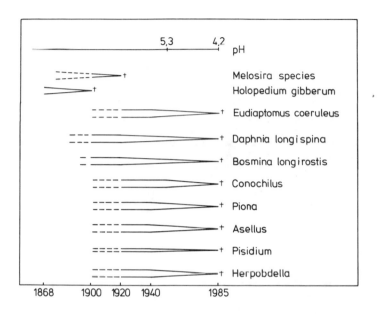

Fig.12. Disappearance of some of the species in Lake Gribsø from 1900
until today.

The bottom fauna was described by Berg (1955),Berg et al.(1956) and
Dunn(1954). Unfinished investigations carried out by Hagelskjær and Olsen
(Freshwater Biological Laboratory, Copenhagen University) have shown that
at least 4 species have disappeared from the lake since 1940: Herpobdella
testacea and Limnophilus nigriceps, and 2 after 1952: Asellus aquaticus
and Pisidium sp. A sediment core with a length of 4.32 metres was
collected from the lake in 1946. The pollen, diatom and chrysophycean
frustules in the core have been studied. The oldest part of the core is
from 5500 B.C. Through an analysis of the diatom shells Nygaard (in Berg
et al.1956) estimated the development of the pH of the lake. About 3000
B.C. the pH was around 7-9, but at 400 B.C. it had dropped to about 5-6,
and 400 A.C. there are indications of a pH as low as 4-5. pH increase
again to about 6 until 1800. In the upper few centimetres of the core the
diatoms indicate a new decrease in pH. This decrease is correlated with
drainage of the catchment area of the lake around 1800. This acidification is still proceeding as clearly shown by recent phytoplankton studies. Phytoplankton samples collected by Wesenberg-Lund in 1900-1919
showed that different species of Melosira disappeared between 1919 and
1926 (Berg et al. 1956). Unpublished investigations of netphytoplankton
from 1920-70 by Nygaard and around 1960 by Kristiansen (Institute of
Thallophyte, Copenhagen University) and samples from 1985 investigated by

Kristiansen have shown further changes.

Today phytoplankton larger than 20 μm are dominated by very few species usually occurring in monocultures e.g. Synura petersenii, S. sphagnicola, Dictyosphaerium pulchella and a species of Chlamydomonas.

Peridinium willei observed from 1907 has not been observed after 1960, but P. inconspicuum has now been recorded. Ceratium has not been recorded in the last years. Three species of Mallomonas recorded regularly since 1913 has not been seen after 1960. Synura petersenii first appeared in 1985. Non motile green algae occurred in numbers in the forties but have disappeared again. The blue green alga Merismopedia has not been recorded since 1952.

Disappearance of species has been recorded from other acidified localities, usually coupled to acidification/oligotrophication and increasing transparency. The acidification of Lake Gribsø has gone further than observed in many Scandinavian lakes. The very acid tributaries to the lake are now under investigations by Pedersen and Sindballe. Their study indicated that the atmospheric deposition of acid materials in the forest is high and is reflected in the water chemistry of the streams. Increasing deposition can explain the accelerating acidification of the lake but changes in land use in the catchment area (fig.3) and windfalls of the spruce in 1968/69 and again in 1981 could also be important factors. According to Freisleben et al.(1986) is the acid deposition on spruce forest in an area 10 kilometers from Lake Gribsø 3.0 kmol ha^{-1} year^{-1}.

Acknowledgements

Thanks to dr. h.c. et scient.G. Nygaard for data, forest supervisor S. Gravsholt, Nødebo Skovdistrikt for housing facilities at the lake and to dr. phil. Claus Nielsen for corrections to the MS.

References

Berg, K. 1955. Ecological remarks on the bottom fauna of a Danish humic acid lake, Store Gribsø. Proc. Int. Ass. Theor. Appl. Limnol. Vol. XII. p. 569-576.

Berg, K. and Petersen I.C.1956. Studies on the humic, acid lake Gribsø Folia Limnol. Scan. no. 8. 273 p. XII tables.

Dunn, D.R. 1954. Notes on the bottom fauna of twelve danish lakes. Vidensk. Medd. fra Dansk Naturh. Foren. bd. 116.251-268.

Freisleben,N.E.v., Ridder C. and Rasmussen, L.,1986, Patterns of acid deposition to a danish spruce forest. Water, Air and Soil Pollution. In Press.

Müller, P.E. 1868. Danmarks Cladocera. - Naturh. Tidsskrift 3 rk. 5 bd. 1868 p. 53 - p. 240.

Nygaard, Gunnar 1965. Hydrographic studies, especially on the carbon-dioxide system, in Grane Langsø. Biol. Skr. Dan. Vid. Selsk:14.2. 110 p.

Nyman, G.H., Oscarson H. G. and Stenson, J. A. E. 1985. Impact of invertebrate predators on the zooplankton composition in acid forest lakes. Ecol. Bull. 37: 239-243.

Rebsdorf, Aage 1983. Are Danish lakes threatened by acid rain? in: Ecological Effects of Acid Deposition. National Swedish Environment Protection Board. Report PM 1636 p. 287-297.

Wesenberg-Lund, C 1904. Plankton Investigations of the danish lakes. Danish Freshwater Biological Laboratory op.3.

Wesenberg-Lund, C. 1940. Gribsø. En dansk dystroph Sø. Naturens Verden 19-34.

CHEMISTRY OF ATMOSPHERIC DEPOSITION AND LAKE ACIDIFICATION IN NORTHERN ITALY, WITH EMPHASIS ON THE ROLE OF AMMONIA

R. MOSELLO, A. PUGNETTI, G.A. TARTARI
CNR Istituto Italiano di Idrobiologia,
Pallanza, Italy

Summary

The only case of a strong acidic lake in North Italy is L. Orta, an important water body located in the L. Maggiore watershed. The acidification is determined by the oxidation of ammonia, discharged in large amounts into the lake by a rayon factory, that has lowered the pH from the original values above 6.0 to the present values of 3.8-4.3 on the whole water column at overturn. The importance of the in-lake processes in causing acidification is confirmed by the fact that the water input from the tributaries is buffered, with a mean total alkalinity of about 0.3 meq/l and a pH generally above 7.0.

The potential acidification capacity of ammonia must be considered also in the case of the atmospheric deposition. Indeed, the data presented in this paper, as well as those of other research, under-line the importance of ammonia among the ionic constituents of rainwater in N. Italy. Even if we take into account the fact that not necessarily all the ammonia is oxidized to nitrate, the contribution of the ammonia transformation in the soil and/or in the lake water to acidification is to be considered of great importance.

INTRODUCTION

The results here presented and discussed, as a contribution to the COST workshop on Reversibility of Acidification, have been obtained as part of a larger research, supported by the E.E.C. and the Italian National Research Council, aimed at studying atmospheric depositions in alpine and subalpine areas of N. W. Italy and the relative effects on subalpine streams and lakes and on high altitude lakes (Mosello, in prep.). This study is the continuation of research activities carried out in this Institute since 1975 both on deposition chemistry (Mosello and Tartari, 1979, 1982, Mosello et al. 1985) and freshwater acidification (Mosello, 1981, 1984 ; Mosello and Tartari, 1983).

In Northern Italy cases of marked freshwater acidification determined by atmospheric deposition are not present, thanks to favourable geological conditions. The only strongly acidic lake is L. Orta, an important water body (surf. 18.2 km^2, z max 143 m, z med 69.3 m) located in the lake Maggiore watershed (Fig. 1). The acidification of L. Orta is due to the oxidation of ammonia, discharged in large amounts into the lake by a rayon factory (Bemberg Company), that has lowered the pH to the present values of 3.8-4.3. Heavy metals (copper, zinc, chromium) deriving from industrial effluents have aggravated the lake conditions. This chemical pollution, as well as the dramatic biological decline of the lake, is described in many papers : Monti, 1930 ; Baldi, 1949 ; Vollenweider, 1963, 1965 ; Bonacina, 1970 ; Bonacina et al., 1973, in press.

The aim of the present contribution is to illustrate two aspects of the research on acidification which are important for a description of present conditions in N. W. Italy : (a) the chemistry of atmospheric deposition in five sites presenting different environmental disturbances and (b) the recovery of L. Orta, strongly acidified by ammonia oxidation, after a drastic reduction of the ammonia input obtained at the end of 1980.

SAMPLING AND METHODS

Atmospheric depositions were sampled during 1985 in five stations (Fig. 1), representing different environmental situations, i.e. : undisturbed high altitude site (L. Toggia, 2160 m a.s.l.) ; subalpine towns (Domodossola and Pallanza, 20000 and 35000 inhabitants, respectively) ; uninhabitated subalpine areas not affected (M. Mesma) and presenting industrial pollution (Bellinzago). The last sampling station is located 3 km W-NW from the thermoelectric plant of Turbigo (1300 MW) and about 20 km from a refinery and a plant producing coal derivatives.

Sampling was carried out with polyethylene bulk collectors, weekly at L. Toggia, Pallanza and Domodossola, every ten days at the other stations. Analyses were performed weekly and monthly respectively for the two groups of stations. The samples collected every ten days for the monthly analyses were filtered and kept in the dark at 4° C to minimize alteration in their ionic composition (Galloway and Likens, 1976 ; Peden and Skowron, 1978).

L. Orta was sampled monthly during 1984 and 1985 at the point of maximum depth. Samples were collected at depths of 0, 5, 10, 20, 30, 50, 100 and 140 m. The tributaries were sampled a total of about 40 times during the two years ; samples were also collected from the discharge of the rayon factory and from the outflow.

Further details on sampling and the analytical methods used are reported in Mosello (in prep) for the atmospheric deposition and in Mosello et al. (in press) for the study of L. Orta.

RESULTS

Atmospheric depositions

The yearly amount of precipitation in 1985 was lower than the pluriannual average in the five sampling stations (Tab. 1) ; the difference is more marked in spring and summer when the volumes of the study year were respectively higher and lower than those of the reference period.

The chemical characteristics of the bulk depositions collected at the five sampling points are reported in Tab. 2 as volume weighted annual means. The mean value of pH was calculated by weighting the hydrogen ion concentrations obtained from the single pH measurement. This method is not strictly correct if there are events with a detectable total alkalinity (Reuss, 1975), resulting in an overestimation of the acidity. Such events are infrequent in the case of the subalpine stations in this area and will be discussed separately. They are more frequent in the Alpine site : for this station and, for comparison, also for Pallanza and Domodossola, the mean volume weighted pH is compared with other statistical indicators of the data distribution, relative to the weekly values (Tab. 3).

The mean annual pH in the four subalpine stations ranges between 4.24 and 4.48 (Bellinzago and Domodossola, respectively) indicating a noticeable level of mineral acidity ; the value for the alpine station (L. Toggia) is 4.91.

As far the other ionic constituents (Tab. 2), ammonium and calcium are the main cations, while sulphate and nitrate are the main anions.

The ratio between sulphate and nitrate ranges between 1.6 and 1.9 in the subalpine stations (M. Mesma and Bellinzago, respectively) ; in the case of the alpine site the ratio is 2.8.

Mean annual conductivity ranges from 43.8 and 22.7 uS/cm at 18° C in the case of the subalpine stations, while it is 10.5 uS/cm at L. Toggia. The highest conductivity was measured at Bellinzago, where the highest concentrations of sulphate, nitrate, ammonia and hydrogen ion were also found. The deposition of the main ions, obtained from the mean annual concentrations and the volume of precipitation, are reported in Tab. 4. Bellinzago presents very high hydrogen ion, ammonium, sulphate and nitrate deposition, close to or higher than the values of Pallanza, though the volume of precipitation of the former is only 64 % of the latter. The lower values of deposition for the subalpine sites were calculated for Domodossola, while the alpine station shows the lowest depositions.

CHEMISTRY OF L. ORTA

Because of the geology of the watershed (geiss, micaschists, granite), the water of the lake was originally poorly buffered, with total alkalinity ranging between 0.3–0.4 meq/l (Monti, 1930). Since 1926 the lake has been subject to a discharge containing ammonium sulphate and copper from a rayon factory ; the in-lake ammonia oxidation exhausted the alkalinity of the waters causing a lowering of pH from the original values of 6-7 down to the present values of 3.8-4.3 and an increase in ammonia and nitrate concentrations (Fig. 2). The copper load from the factory was drastically reduced in 1956, and an ammonium sulphate recovery plant came into operation in 1980, reducing the nitrogen load from 1950-3350 t N/a (years 1958-1979) to about 30 t N/a (Bonacina et al., in press). The response of the lake, whose chemical condition was quite stable in the period 1970-1982 (Fig. 2), became apparent in 1984, aided also by the heavy rain which characterized the spring of that year (Mosello et al., in press). Tab. 5 shows the chemical characteristics of the lake waters at the overturn in 1984 and 1985. The main differences between the two years are the strong decrease in ammonia (from 284 to 164 ueq/l) and the increase in hydrogen ion (from 44 to 126 ueq/l). The seasonal variations in oxygen, pH, ammonia and nitrate concentrations (Fig. 3) emphasize the relationship between ammonia oxidation and pH decrease.

Oxygen, which in March displays concentrations of 8 mg/l on the whole water column, begins to show a sharp decrease in the hypolimnion in July-August, until complete anoxia below a depth of 100 m is reached in December-January. pH, with a value of 4.40 at the 1984 circulation, maintains values of 4.4-4.5 in hypolimnetic waters until August ; in the following months there is a sharp decrease, leading to the lowest values of 3.8-3.9 on the whole water column in January-February 1985.

This decrease in pH must be connected to the decrease in ammonia, which also begins in August ; the values of the latter fall from 270 to 190 ueq/l, while the corresponding increase in nitrate is more modest (from 290 to 340 ueq/l). This difference may be explained by the output of ammonia and nitrate through the outlet, as shown by mass balance calculation (Mosello et al., in press).

DISCUSSION AND CONCLUSION

Many different redox reactions, occurring both in the watershed and in the lake, may influence the alkalinity/acidity budget (see, e.g., Stumm and Morgan, 1981 ; Van Breemen, Driscoll and Mulder, 1984). In the case of L. Orta, where the ammonia load was exceedingly higher than is usual in this area, acidification deriving from the process of in-lake oxidation to nitrate is particularly evident. This process has been recognized since

1961 as the cause of the lake's acidification (Vollenweider, 1963) ; the same conclusion results from a study on the chemical budget of the major ions carried out considering the main tributaries, the atmospheric input and the outflow (Mosello et al., in press). Ammonia oxidation has been recognized by other authors as a source of acidity in lake ecosystems (Kelly et al., 1982 ; Schindler et al., 1985), and in the soil (Reuss, 1975 ; Ulrich, 1983 ; Van Breemen et al., 1984 ; Matzner and Ulrich, 1985).

In areas with high ammonia emissions, such as The Netherlands, Belgium and Denmark (Buijsman et al., 1985) ammonia oxidation is considered one of the main causes of acidification (Schuurkes, 1986 ; Ministry of Housing, Physical Planning and Environment, 1986).

The ionic composition of atmospheric deposition at the five sampling stations considered in this paper shows high ammonia concentration also for the Italian subalpine station (Tab. 2) ; even higher concentrations have been measured in sampling sites located in the Po Valley (Gruppo di Studio, 1985). Indeed, N. Italy is indicated as an area of high ammonia emission, on the basis of livestock and industrial activities (Buijsman et al., 1985). Furthermore, the abundant rainfall that is a feature of the subalpine area determines a high ammonia load to the soil (Tab. 4).

These observations indicates that it is incorrect to consider the hydrogen ion load deriving from the atmosphere as the only acidifying factor , in the case of N. Italy this may result in a strong underevaluation of the acid load. The maximum potential acidifying capacity may be evaluated summing the hydrogen ion load with twice the ammonia load, taking into account the stoichiometric ratio NH_4^+ / H^+ = 1/2 of the ammonia oxidation to nitrate. The actual acidifying capacity might be lower than this value, since not necessarily all the ammonia is completely oxidized to nitrate ; indeed, ammonia may undergo several other chemical processes, depending on the main characteristics of the soil considered (geology, pedology, land use, removal of biomass in or from the system). We think that more detailed studies on these aspects of soil and freshwater acidification are needed in the near future, in the framework of a more general approach to the acidification problem. Both the case of L. Orta and the ammonia concentrations in atmospheric deposition in N. Italy stress the importance of these topics for this part of the Country.

ACKNOWLEDGEMENT

We are grateful to our colleagues and technicians at the C.N.R. Istituto Italiano di Idrobiologia for their co-operation at all stages of the work. We also thank Dr. J.A.A.R. Schuurkes, of the Catholic University, Toernooiveld, Nijmegen, The Netherlands, for the useful exchange of opinions on the topics examined in the paper.

REFERENCES

BALDI, E. 1949. Il Lago d'Orta, suo declino biologico e condizioni attuali. Mem. Ist. Ital. Idrobiol., 5 : 145-188.

BONACINA, C. 1970. Il Lago d'Orta : ulteriore evoluzione della situazione chimica e della struttura della biocenosi planctonica. Mem. Ist. Ital. Idrobiol., 26 : 141-204.

BONACINA, C., G. BONOMI and R. MOSELLO. in press. Advances in the recovery of Lake Orta after three years of functioning of the new treatment plants. Mem. Ist. Ital. Idrobiol., 44.

BONACINA, C., G. BONOMI and D. RUGGIU. 1973. Reduction of the industrial pollution of Lake Orta (North Italy) : an attempt to evaluate its consequences. Mem. Ist. Ital. Idrobiol., 30 : 149-168.

BUIJSMAN, E., J.F.M. MAAS and W.A.H. ASMAN. 1985. Ammonia emission in Europe. IMOU report R-85-2. Inst. Meteorology and Oceanography State University, Utrecht, The Netherlands.

GALLOWAY, J.N. and G.E. LIKENS. 1976. Calibration of collection procedures for the determination of precipitation chemistry. Water Air Soil Poll., 6 : 241-258.

Gruppo di Studio sulle caratteristiche chimiche delle precipitazioni. 1985. Deposizioni atmosferiche sul Nord Italia (Ottobre 1982 - Settembre 1983). Acqua Aria, 8 : 721-735.

KELLY, C.A., J.W.M. RUDD, R.B. COOK and D.W. SCHINDLER. 1982. The potential importance of bacterial processes in regulating rate of lake acidification. Limnol. Oceanogr., 27 : 868-882.

MATZNER, E. and B. ULRICH. 1985. Implications of the chemical soil conditions for forest decline. Experientia, 41 : 578-584.

Ministry of Housing, Physical Planning and Environment. 1986. The role of ammonia in acidification in the Netherlands. 8 pp.

MONTI, R. 1930. La graduale estinzione della vita nel Lago d'Orta. Rend. R. Ist. Lomb. Sc. Lett., 63 : 3-22.

MOSELLO, R. 1981. Chemical characteristics of fifty Italian alpine lakes (Pennine-Lepontine Alps), with emphasis on the acidification problem. Mem. Ist. Ital. Idrobiol., 39 : 99-118.

MOSELLO, R. 1984. Hydrochemistry of high altitude alpine lakes. Schweiz. Z. Hydrol., 46 : 86-99.

MOSELLO, R. in prep. Effects of acid precipitation on lake ecosystems in North-Western Italy. Final Report CEE-CNR ENV. 878 I (S).

MOSELLO, R., C. BONACINA, A. CAROLLO, V. LIBERA and G.A. TARTARI. in press. Acidification due to in-lake ammonia oxidation : an attempt to quantify the proton production in a highly polluted subalpine Italian lake. Mem. Ist. Ital. Idrobiol., 44.

MOSELLO, R. and G. TARTARI. pH e caratteristiche chimiche delle acque meteoriche di Pallanza. Mem. Ist. Ital. Idrobiol., 37 : 41-49.

MOSELLO, R. and G. TARTARI. 1982. Chemistry of the precipitation in the L. Maggiore watershed (N. Italy). Mem. Ist. Ital. Idrobiol., 40 : 163-180.

MOSELLO, R. and G. TARTARI. 1983. Effects of acid precipitation on subalpine and alpine lakes. Water Quality Bulletin, 8 : 96-100.

MOSELLO, R., G. TARTARI and G.A. TARTARI. 1985. Chemistry of bulk deposition at Pallanza (N. Italy) during the decade 1975-84. Mem. Ist. Ital. Idrobiol., 43 : 311-322.

PEDEN, M.E. and L.M. SKOWRON. 1978. Ionic stability of precipitation samples. Atmospheric Environment, 12 : 2343-2349.

REUSS, J.O. 1975. Chemical/biological relationships relevant to ecological effects of acid rainfall. EPA - 660/3-75-032, 46 pp.

SCHINDLER, D.W., M.A. TURNER and R.H. HESSLEIN. 1985. Acidification and alkalinization of lakes by experimental addition of nitrogen compounds. Biogeochemistry, 1 : 117-133.

SCHUURKES, J.A.A.R. 1986. Atmospheric deposition and its role in the acidification and nitrogen enrichment of poorly buffered aquatic systems. Experientia, 42 : 351-357.

STUMM, W. and J.J. MORGAN. 1981. Aquatic chemistry. An introduction emphasizing chemical equilibria in natural waters. Wiley and Sons, New York, 780 pp.

ULRICH, B. 1983. Soil acidity and its relations to acid deposition. In : B. ULRICH and J. PANKRATH (eds.), Effects of accumulation of air pollutants in forest ecosystems. Reidel Publ. : 127-146.

VAN BREEMEN, N., C.T. DRISCOLL and J. MULDER. 1984. Acidic deposition and internal proton sources in acidification of soil and waters. Nature, 307 : 599-604.

VOLLENWEIDER, R.A. 1963. Studi sulla situazione attuale del regime chimico e biologico del Lago d'Orta. Mem. Ist. Ital. Idrobiol., 16 : 21-125.

VOLLENWEIDER, R.A. 1965. Materiali ed idee per una idrochimica delle acque insubriche. Mem. Ist. Ital. Idrobiol., 19 : 213-286.

Tab. 1 - Amount of precipitation during the sampling period and historical
means (mm/a)

	Year 1985	Reference period
Pallanza	1517	1721 (1951-1984)
Bellinzago	969	918 (1921-1950)*
M. Mesma	1233	1781 (1921-1970)
Domodossola	1270	1378 (1921-1984)
L. Toggia	860	1278 (1951-1984)

* Value measured at Novara, about 30 km from the sampling station

Tab. 2 - pH, volume weighted annual means of the main ions (ueq/l) and
conductivity (uS/cm at 18 C) in bulk deposition

	pH	H+	NH4+	Ca++	Mg++	Na+	K+	Cat.	SO4--	NO3-	Cl-	Br-	An.	Ions	Cond
Pallanza	4.43	37	45	48	9	25	3	167	77	44	10	2	133	300	28.5
Bellinzango	4.24	58	66	40	10	15	5	194	126	66	20	2	214	408	43.8
M. Mesma	4.42	38	57	26	6	9	5	141	80	50	12	0	142	283	29.2
Domodossola	4.48	33	28	30	7	6	2	106	59	33	7	1	100	206	22.7
L. Toggia	4.91	12	15	21	6	10	2	66	37	13	6	0	56	122	10.5

Tab. 3 - Statistical indicators of the distribution of the values of pH
compared with the volume weighted means (%le = percentile,
n=number of data)

	min	25%le	50%le	75%le	max	mean	n
Pallanza	3.86	4.10	4.35	4.59	6.84	4.43	32
Domodossola	3.87	4.24	4.41	4.58	6.22	4.48	32
L. Toggia	4.29	4.80	5.05	5.39	6.25	4.91	30

Tab. 4 – Deposition of the main ions (meq/m^2 a)

	H+	NH4+	Ca++	Mg++	Na+	K+	Cat.	SO4--	NO3-	Cl-	Br-	An.	Ions
Pallanza	56	68	73	14	38	5	254	117	67	15	3	202	456
Bellinzago	56	64	39	10	15	5	189	122	64	19	2	207	396
M. Mesma	47	70	32	7	11	6	173	99	62	15	0	176	349
Domodossola	42	36	38	9	8	3	136	75	42	9	1	127	263
L. Toggia	9	11	15	4	7	1	47	27	10	4	0	41	88

Tab. 5 – Chemical characteristics of L. Orta at the overturn (January 1984 and 1985). Units in ueq/l, conductivity in uS/cm at 18 'C

	Jan 1984	Jan 1985
pH	4.36	3.90
H+	44	126
NH4+	284	164
Ca++	315	324
Mg++	123	123
Na+	238	248
K+	27	28
SCat	1031	1013
SO4--	641	645
NO3-	76	66
Cl-	319	321
SAn	1036	1032
Sions	2067	2045
Cond	124	153

93

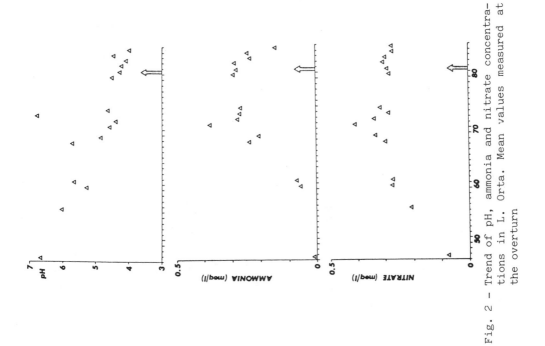

Fig. 2 - Trend of pH, ammonia and nitrate concentrations in L. Orta. Mean values measured at the overturn

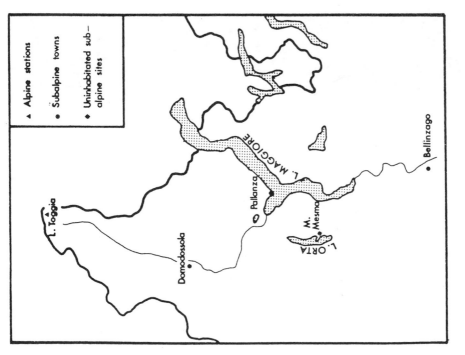

Fig. 1 - L. Orta and geographical distribution of the sampling stations of atmospheric depositions

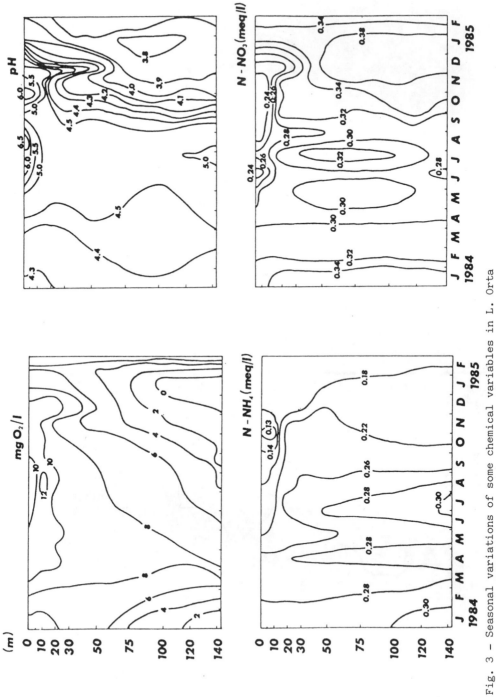

Fig. 3 – Seasonal variations of some chemical variables in L. Orta

BUFFERING MECHANISMS IN ACIDIFIED ALPINE LAKES

Jürg Zobrist, Laura Sigg, Jerald L. Schnoor* and Werner Stumm

Swiss Federal Institute for Water Resources and Water
Pollution Control (EAWAG), CH-8600 Dübendorf,
Swiss Federal Institute of Technology (ETH), CH-8093 Zürich (Switzerland)
*University of Iowa, Iowa City, Iowa 52242, U.S.A.

Summary

Under the climatological conditions prevailing in alpine regions,
chemical weathering of rocks represents the principal reaction that
neutralizes the acid inputs into the watersheds and that affects the
water composition. The load of strong acids is due to wet and dry
atmospheric deposition and is also due to the oxidation and assimilation of the ammonium, that is inserted by atmospheric deposition
as well.
4 small and shallow lakes have been studied during a snow-free period. They are situated in the central part of the Alpes above 2000 m
in an area with granitic gneiss and little soil. The lakes not influenced by neighbouring calcareous rocks have a pH between 4.9 and
6.0, e.g. the alkalinity varies between -10 and + 20 μequiv/ℓ. The
atmospheric input of strong acids amounts to $\sim$ 15 mequiv/m^2 y, the
same quantity is additionally produced by ammonium oxidation and
assimilation. The weathering rate of the cristalline rocks is
$\sim$ 20 mequiv/m^2 y.

1. INTRODUCTION

The occurrence of acid deposition results from the anthropogenic
disturbance of many cycles of elements which couple the atmosphere with
the hydrosphere and the pedosphere (1). As a consequence of the fossil
fuel combustion, the natural redox conditions in the atmosphere have
changed. Nowadays, the oxidation rates of carbon, sulfur and nitrogen
exceed the corresponding reduction rates. This disequilibrium gives rise
to an enhanced production of hydrogen ions in the atmosphere.

The impact of acid atmospheric deposition into surface waters in
Scandinavia, in the eastern part of the United States and Canada has been
well documented (2). Less known are the acidified lakes in Central Europe
(3) and in the alpine areas (4,5).

Small alpine lakes, situated above the timberline where there is
little soil, represent a simple natural system for studying the interaction between acid atmospheric deposition and cristalline bedrocks.

In this article, we shall discuss the processes that govern the neutralization of acid deposition in alpine lakes and which regulate the
water composition.

2. STUDY AREA

The four lakes studied are situated in the upper Maggia valley in the southern central part of the Alps, see Fig. 1. The small and shallow lakes are located well above the timberline, in an area with cristalline rocks that is only partially covered with a thin layer of soil.

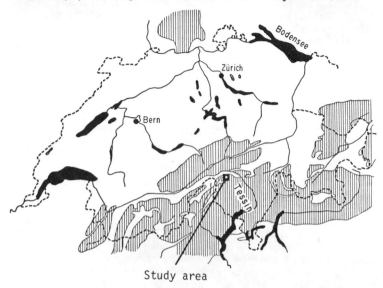

Study area

Figure 1   Map of Switzerland, displaying the main surface waters and the areas with cristalline rocks (lined area).

As a consequence of the high altitude, more than 50% of the wet deposition (1.6 to 2 m/y) is precipitated as snow. Therefore, the run-off occurs predominantly during the summer months. For three lakes, the retention time of the water is very short, e.g. in the order of a few days, see Table 1. The values indicated represent the average over the run-off season (5 months for the lower two lakes, 4 months for the others). They may fluctuate by a factor of 3 during the run-off period.

Table 1   Characteristics of the lakes studied

Lake	Altitude m	Lake Area ha	Drainage Area ha	Retention Time d
Inferiore	2070	5.0	165	~ 9
Superiore	2130	6.9	105	~ 40
Piccolo Naret	2350	2.1	125	~ 2
Cristallina	2400	0.75	17	~ 5

The bedrock in the drainage areas consists of granitic gneiss with a mineral composition of quartz (~ 30%), plagioclase, mainly anorthite (~ 25%), K-feldspar (~ 20%), biotite (~ 15%) and epidote (~ 10%) (6). However, in the northern adjacent basins, schist (Bündner Schiefer) is the predominant rock which contains calcareous minerals.

3. BUFFERING MECHANISMS IN NATURAL WATERS

3.1 ACID AND BASE-NEUTRALIZING CAPACITIES

Considering the regulation of the pH in natural waters, one needs to distinguish between the hydrogen-ion concentration (or activity) as an intensity factor, and the availability of protons, the hydrogen-ion reservoir. This capacity parameter is given by the base neutralizing capacity, also called strong acidity [H-Acy] or just acidity. The strong acidity may be defined as the sum of the concentrations of all the species containing protons in excess, minus the concentration of the species containing protons in deficiency in respect to the proton reference level. For natural waters a convenient reference level is H_2O and H_2CO_3 (7):

$$[\text{H-Acy}] = [H^+] - [HCO_3^-] - 2[CO_3^{2-}] - [OH^-]$$

For natural water with pH <6:

$$[\text{H-Acy}] = [H^+] - [HCO_3^-]$$

In waters with a low ionic content other acids, such as organic acids (HA) with pK_a values $\leqslant 6$ and aluminium-hydroxo species, have to be included too:

$$[\text{H-Acy}] = [H^+] + [HA] + \underbrace{3[Al^{3+}] + 2[Al(OH)^{2+}] + [Al(OH)_2^+]}_{\nu \cdot Al^{\nu+}, \text{ where } \nu \text{ depends on the pH}} - [HCO_3^-]$$

This equation may also be used in order to define natural acid waters, namely [H-Acy] >0.

Considering the charge balance of the water, the acidity may alternatively be defined as:

$$[\text{H-Acy}] = 2[SO_4^{2-}] + [NO_3^-] + [Cl^-] + [A^-]$$
$$- 2[Ca^{2+}] - 2[Mg^{2+}] - [Na^+] - [K^+] - [NH_4^+]$$
$$= \sum[\text{conservative anions}] - \sum[\text{base cations}]$$

This means that any changes in the concentration of cations or anions cause a shift in acidity, e.g. if rain water is percolating through soil that only adsorbs sulfate [and releases OH-ions], the acidity will decrease.

In an analog manner, the acid-neutralizing capacity called alkalinity (Alk) is defined as:

$$[\text{Alk}] = -[\text{H-Acy}] = [HCO_3^-] + 2[CO_3^{2-}] + [A^-] + [OH^-] - [H^+]$$

3.2 PROTON BALANCE

The acidity or alkalinity measured in natural waters represents the disparity between the processes that add or produce acidity and those consuming protons (8). These processes modifying the proton balance are listed in Table 2.

The weathering reactions of silicate minerals cannot be formulated unambigously, since the dissolution reaction is mostly incongruent and different reaction products are possible. In any case, there exists a 1:1 relationship (expressed in equivalents) between the protons consumed and the base cations produced. Ion-exchange reactions are limited by the ion-exchange capacities of the solid phases. The exchangeable ions have to be replaced by an additional chemical reaction, e.g. base cations are provided by weathering. The synthesis of biomass represents a quasi "amphoteric" reaction; if more base cations than conservative anions are incorporated in the biomass, the acidity will increase. All the processes quoted in Table 2 take place in the watershed as well as in the water-body and its sediments. The importance of the individual processes depends on the occurrence of the corresponding reactants, the redox conditions and the hydrological conditions.

Table 2   Processes which modify the proton balance

Processes supplying protons: Quantity:

Atmospheric depositions:

- wet deposition   the contribution of these pro-
- dry deposition of SO_2 with   cesses has to be measured or
  consecutive oxidation and of   estimated
  other acidic species

                                              Change in acidity
                                              (equivalents per mole reacted,
                                              reactant is underlined)

Oxidation reactions:

- Nitrification

  $\underline{NH_4}^+ + 2\ O_2 \rightarrow NO_3^- + 2\ H^+ + H_2O$ 2

- Sulfide oxidation

  $\underline{HS}^- + 2\ O_2 \rightarrow SO_4^{2+} + H^+$ 1

Ammonium assimilation:

  $a\ CO_2 + b\ H_2O + c\ \underline{NH_4}^+$
  $\rightarrow (CH_2O)_b\ (NH_3)_c + c\ H^+ + a\ O_2$ c

Processes consuming protons:

Weathering:

- Carbonate minerals

  $\underline{CaCO_3}$ or $\underline{MgCO_3} + 2\ H^+ \rightarrow Ca^{2+}$ or $Mg^{2+} + H_2CO_3$ -2

- Silicate minerals,

  primary rock minerals + $H^+ + H_2O$

  $\rightarrow$ cations + H_4SiO_4 + weathered minerals (secondary),

  e.g. $7\ \underline{NaAlSi_3O_8} + 6\ H^+ + 20\ H_2O$ -6/7
  $\rightarrow 6\ Na^+ + 10\ H_4SiO_4 + 3\ [Na_{1/3}\ Al_{2\ 1/3}\ Si_{3\ 2/3}\ O_{10}(OH)_2]$

Ion exchange:

  $2\ R\ OH + \underline{SO_4}^{2+} \rightarrow R_2SO_4 + 2\ OH^-$ -2
  $NaR + \underline{H}^+ \rightarrow HR + Na^+$ -1

Reduction Reactions:

- Denitrification

  $\underline{NO_3}^- + \frac{5}{4}\ CH_2O + H^+ \rightarrow \frac{1}{2}\ N_2 + \frac{5}{4}\ H_2CO_3 + \frac{1}{2}\ H_2O$ -1

- Sulfate reduction

  $\underline{SO_4}^{2-} + 2\ CH_2O + 2\ H^+ \rightarrow H_2S + 2\ H_2CO_3$ -2

- Iron oxide reduction

  $\underline{FeOOH} + \frac{1}{4}\ CH_2O + H^+ \rightarrow Fe^{2+} + \frac{1}{4}\ H_2CO_3 + OH^- + \frac{1}{2}\ H_2O$ -1

"Amphoteric" processes:

Production of biomass

  $a\ CO_2 + b\ \underline{NO_3}^- + c\ \underline{NH_4}^+ + d\ \underline{HPO_4}^{2-} + e\ (\underline{SO_4}^{2-})$
  $+ f\ \underline{Ca^{2+}} + g\ \underline{K^+} + .. + m\ H_2O + (b-c+2d+2e-2f-g..)H^+$ [c+2f+g+...
  $\rightarrow [(CH_2O)_a(NH_3)_{b+c}(H_3PO_4)_d(H_2SO_4)_e(CaO)_f(K_2O)_g(H_2O)_m]$ -b-2d-2e]
  $+ (a+2b)O_2$

In an alpine area with less soil and a small biological productivity, protons are mainly supplied by atmospheric deposition and oxidation reactions. Weathering of rocks represents the important process consuming hydrogen-ions. Under the conditions stated, the contribution of the ion-exchanges may be minor only.

4. RESULTS

The data gained from the sampling campaign during 1985 are represented in Fig. 2. The total ionic contents in these clearwater lakes are small, also in comparison to the majority of the Scandinavian surface waters (2).

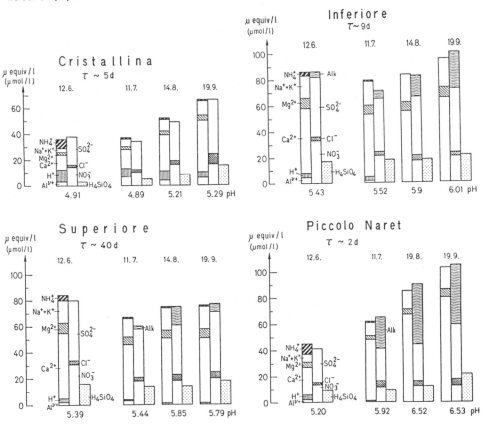

Figure 2   Evolution of the concentrations in four alpine lakes during summer 1985.

In mid-June, the upper two of the lakes, Cristallina and Piccolo Naret, which are very shallow, were still filled with melting snow. In addition, these lakes had a minor outflow. Therefore, the water composition measured represents the composition of the bulk precipitation (snow and dry deposition) fallen during the preceeding winter. Indeed, the concentrations measured in these two lakes agree with those obtained in the bulk precipitation at the Italian sampling station Lago Toggia (10 km WSW of Lake Cristallina, altitude 2160 m (9)).

The evolution of the water composition during summer has to be interpreted in connection with the retention time of the lakewater. In Lake Cristallina and Piccolo Naret the water renewal takes a couple of days. Therefore, the changes in concentrations reflect the chemical processes occurring in the drainage area. The distinct increase in silicic acid and in the base cations imply the weathering of cristalline rocks. The remarkable augmentation of the calcium and bicarbonate concentration is due to a dissolution of calcite. The disappearance of the ammonium is caused by the nitrification and the ammonium assimilation. In Lake Superiore, characterized by a retention time of about 40 days, the water composition stays nearly as constant in summer as during the whole year. The small and proportional decrease in the July concentration of the main constituent is caused by the dilution with snow water. In addition, the total ionic content is about twice as high as that in the bulk precipitation. Therefore, the water composition in the lake reflects the steady state condition between the various processes taking place in the drainage area and in this lake.

5. MASS BALANCE CALCULATIONS

The mass fluxes regarding the major components in the outflows of the lakes can be calculated from our data by assessing the average water flow. The atmospheric input, however, has to be estimated from bulk precipitation measurements at Lago Toggia (9) and from the water composition in the upper two lakes.

Table 3   Yearly mass fluxes of the major components in the catchment area of Lake Superiore

Constituent	Input atmosphere snow, rain, dry*, $\sum$				Outflow Lake	Difference outflow - input
water	1.0	0.6$[m]$	-	1.6	$\begin{bmatrix}m\end{bmatrix}$ 1.0	$\begin{bmatrix}m\end{bmatrix}$ 0.6
	{m equiv/m^2}				{m equiv/m^2}	{m equiv/m^2}
NO_3^-	9	9	-	18	22	4
Cl^-	2	3		5	4	1
SO_4^{2-}	13	23	8	44	42	-2
HCO_3^-	-	-		-	6	6
H^+	4	2	8	14	1	-13
Ca^{2+}	12	22	-	34	48	14
Mg^{2+}	2.0	1.8	-	3.8	7.0	3.2
Na^+	2.0	2.4	-	4.4	8.5	4.1
K^+	1.0	1.8	-	2.8	7.8	5.0
NH_4^+	5	6		11	2	-9
Al^{v+}				<0.2	2.5	2.5
	m mol/m^2					
H_4SiO_4				<0.2	16	16

*In the alpine area, the yearly average of the SO_2 concentration in the air is ~ 2 µg SO_2/m^3, a deposition velocity of $4 \cdot 10^{-3}$ m/s is assumed.

The results of the calculation (an example is given in Table 3) show the following common characteristics:
- For the conservative anions chloride and sulfate, the massflux in the outflow counterbalances approximately the atmospheric input.
- Hydrogen- and ammonium-ions are retained in the drainage area.
- There is a net production of base cations, of silicic acid, of nitrate and of bicarbonate in the watershed.

These differences in the mass fluxes between the atmospheric input and the outflow can be attributed to the processes discussed hereafter:
- Weathering of silicate rocks:
  The following reaction equation may be assumed by applying the general weathering reaction on the minerals which are present in the basins

$$[(Ca, Mg, Na, K)_x (SiO_2)_y (AlO_{3 or 2})_z] + (1+cv)H^+ + 2bH_2O$$
$$\rightarrow (Ca, Mg, Na, K)^+ + b\ H_4SiO_4 + c(Al^{v+})$$
$$+ [(Ca, Mg, Na, K)_{x-1} (SiO_2)_{y-b} (AlO_{3 or 2})_{z-c}]$$

  whereby: $b \sim 1$, $c \sim 0.1$

  This assumption for the coefficient b implies that the quantity of base cations (expressed in equivalents) is equal to the amount of silicic acid (expressed in mols) produced and to the amount of silicates weathered.
- Weathering of calcite:
  Although the geological map does not show any calcareous rock in the catchment area of the four lakes, a dissolution of calcite has also to be taken into consideration. The production of bicarbonate and calcium in three of the four drainage areas is larger than expected for the weathering of anorthite. The occurrence of calcareous minerals could be due to tiny inclusions and/or to wind-transported calcareous dust.
- Nitrification and ammonium assimilation:
  The mass balance calculations suggest that only a part of the ammonium is transformed into nitrate.

Our data do not indicate any significant contribution of the other processes cited in Table 2.

Table 4 summarizes the reaction rates of the occurring processes and presents the acidity budget.

The reaction rates were obtained by using the differences in the mass fluxes between the outflow and the atmospheric deposition in Table 3 as input into the stoichiometric calculations with the reactions listed in Table 2. The calculation of the weathering rate of silicates does not consider the sedimentation rate of diatoms which represents a sink for silicic acid. This rate can be assumed to be less than 5 m mol Si/m^2 y.

The interpretations of Table 4 are included in the conclusions. The agreement of the calculated and measured changes in acidity indicates that the important processes have been considered.

Table 4   Reaction rates and acidity budget of the processes occurring in the four drainage basins

Process	Drainage area of lake:			
	Inferiore	Superiore	P. Naret	Cristallina
Reaction rates $[m\ equiv/m^2\ y]$				
Weathering of silicates	19	16	18	16
Weathering of calcite	19	6	45	0
Nitrification	5	4	3	0
NH_4–Assimilation	5	5	9	12
Acidity (proton) budget $[m\ equiv/m^2y]$				
Increase by: (proton supply)				
-Atmospheric deposition	14	14	17	17
-Nitrification	10	8	6	0
-NH_4 Assimilation	5	5	9	12
Total supply	29	27	32	29
Decrease by: (proton consumption)				
-Silicate weathering	21	18.5	19	23
-Calcite weathering	19	6	45	0
Total consumption	40	24.5	64	23
Change in acidity:				
-calculated	-11	2.5	-32	6
-measured	-12	-4	-54	8

6. CONCLUSIONS

The following conclusions can be drawn with respect to an alpine cristalline region covered with little soil:
- The weathering reactions represent the principal process that neutralizes the strong acids supplied by atmospheric deposition and produced in the system.
- The transformation reactions of ammonium also contribute significantly to the input of strong acids. The ammonium also originates from atmospheric deposition.
- The weathering rate of the granitic gneiss is small and amounts to about 20 m equiv/m^2 per year (= 200 equiv/ha y).
- The mass flux of the major lake water constituents from the atmosphere into the watershead exceeds that from silicate weathering.
- Adjacent basins with calcareous bedrock may influence the composition of surface waters. This effect may be due to small calcareous inclusions or to wind-transported dust.
- Oxidation and reduction processes occurring in the watershed may significantly influence the acidity budget of the aquatic system.

7. ACKNOWLEDGEMENT

The work reported has been supported by the Swiss National Foundation (National program 14 on air quality). We thank Fritz Zürcher, David Kistler, Claudia Maeder and Ursula Michel for their chemical analytical work and Claude Jaques for his technical assistance.

REFERENCES

(1) Stumm W., Morgan J.J., Schnoor J.L. (1983). Saurer Regen, eine Folge der Störung hydrogeochemischer Kreisläufe. Naturwissenschaften 70, 216-223.

(2) Drables D. and Tollan A. eds. (1980). Ecological impact of acid precipitation. Proc. Symp. SNSF Projekt, Oslo.

(3) Paces T. (1985). Sources of acidification in Central Europe estimated from elemental budgets in small basins. Nature 315, 31-36.

(4) Stumm W., Righetti G. (1982). Tessiner Bergseen: saurer Regen, saure Traufe. Neue Zürcher Zeitung, Beilage Forschung und Technik, Fernausgabe Nr. 231, S. 31.

(5) Mosello R., Tartari G. (1983). Effects of acid precipitation on subalpine and alpine lakes. Water Quality Bulletin 1983, 96-100.

(6) Simpson C. Dissertation ETH Zürich (1981).

(7) Stumm W., Morgan J.J. (1981). Aquatic Chemistry, 2nd. ed., Wiley-Interscience, New York, 163-210.

(8) Stumm W., Schnoor J.L. (1985). Acidification of aquatic and terrestrial systems. In: Chemical Processes in Lakes, W. Stumm ed., Wiley-Interscience New York. 311-338.

(9) Mosello R. Private communication.

ACIDITY MITIGATION IN A SMALL UPLAND LAKE

Gwyneth Howells
Department of Applied Biology
University of Cambridge

SUMMARY

At a small upland acid lake in S.W. Scotland data on deposition, catchment characteristics and water quality will be used to estimate the changes that are likely to result from alternative mitigation strategies. These might include changes in deposition, land or vegetation management, or lime applications.

1. INTRODUCTION

Loch Fleet in Galloway, south west Scotland, is a small (17 ha) lake (Fig. 1) receiving drainage from a 107 ha catchment lying on coarse muscovite-biotite granite. This bedrock is overlain by peaty soils (mean depth about 30 cm) containing little mineral till and a large accumulation of organic material. The soils are acid (pH 2.5 to 3.5, 0.1 M $CaCl2$) and are strongly gleyed and leached (Bown and Heslop, 1979).

The altitude of the lake is 344 m and rainfall is in excess of 2000 mm each year. The site is exposed to prevailing winds from the west and southwest. Rain acidity has a pH value of 4.83 (1985 weighted mean) equivalent to 15 µeq H/l and rainfall in 1985 was 2153 mm, 101% of the 30 year mean for this area. The lake is acid (generally pH 4.0 to 4.5) and is reported to have been less acid in the 1960's (pH 5.5 to 6.0). Diatom remnants in the lake sediments suggest that acidification is quite recent, since about 1970 (Batterbee, pers. comm.).

Fish (brown trout) were present in the lake until about 1970. A catch of about 100 fish/year was reported until 1955, then declined sharply over the following decade although one or more fish were reported after 1960 (Cally Estate records). Stocking was never a regular practice, so this population was self-maintained as upstream immigration is precluded by a 5 m fall at Craigie Linn, about 7 km downstream. Present water quality is considered unsuitable for trout (typically pH 4.0, calcium 1 mg/l, aluminium 100 µg/l) and test fishings in 1984 failed to find fish in the lake, or trout in the downstream reach. Eels, however, are present in the stream below the lake and both eels and trout are present below the fall.

On the western shore, an area (12% of the total catchment) was afforested in 1961; the species planted were Sitka spruce and lodgepole pine, with some larch along the lake shore. The remainder of the catchment consists of rough moorland dominated by Molinia caerulea (flying bent) or Calluna vulgaris (heather). In some sectors of the catchment there are quite extensive areas of Sphagnum; such boggy areas are a significant portion of sector VII, feeding the principal inflow stream. The general features of the site are summarised in Table 1.

Prior to the area being taken over for forestry in the 1960's, it had been the practice to burn the moorland on a regular basis to improve the grazing for sheep and goats by control of <u>Molinia</u>, and to improve the heather on which grouse are dependent. Burning regimes were practiced in the area from the end of the 18th century until about 1955, with a systematic and comprehensive "strip burning" regime established by 1870 and continuing through until the 1950's (NCC, 1983).

Present water quality is unsuitable for the local strain of trout although other strains are not so severely affected. In addition, the present potential for spawning is limited or negligible due to the absence of suitable gravels and the very variable flows in the inlet stream. However, past records of spawning and resident fish suggested that the lake was a suitable site for an intervention programme aimed at restoring the fishery. A five-year project was initiated in 1984 by the CEGB with the support of the Scottish Boards (SSEB and NSHB) and the National Coal Board (NCB). The objectives were:

(a) to demonstrate that the water chemistry of a lake can be brought to a range suitable for trout by one or more of several treatments to the catchment or waters, manipulation of hydraulic contact with natural or other basic minerals, manipulation of the ion exchange system, e.g. by burning;

(b) once suitable chemical conditions have been achieved in portions of the lake, to demonstrate the suitability of the water for a self-sustaining trout population.

The project is further expected to provide knowledge and experience of practical land, lake and stream management practices which can be applied to a range of site conditions, and so will be of general application.

The project, within its limited scope, timescale and resources, is not intended as a study of acidification and its effects, nor is it strictly a catchment budget study. None the less, it is likely that the quantitative information gained, and the increased knowledge of natural processes in this acidified system will further our understanding of acidification and its control or mitigation. In this paper the data will be used to compare the effectiveness of emission reduction, vegetation management, and lime applications as alternative options for the improvement of water quality.

2. APPROACH

The project is designed to follow a sequence of steps necessary to test alternative manipulative techniques with regard to their practicality, cost and efficacy in improving water quality. Rather than select a number of apparently "similar" lakes in this region, treatments were to be applied to independent sectors of the lake catchment, with the consequent effects monitored in vegetation, soil, soil percolates, runoff, and portions of the lake itself. As water quality is improved, it would be tested against brown trout survival both in acute exposures and the long term. In parallel, the spawning potential and productivity of the lake would be assessed and improvements made if considered necessary. When the most effective and economic treatment(s) is(are) identified, the whole catchment would be treated and the lake restocked with fish.

The project thus falls into 4 phases;

(1) an initial phase of collecting and collating baseline information;

(2) a second phase when sections of the lake are isolated as "embayments", and weirs and "flow-boxes" installed to provide the means of sampling drainage waters qualitatively and quantitatively;

(3)   a third phase of manipulative interventions to improve water quality, and of monitoring these changes;
(4)   a final phase for overall catchment treatment and fish restocking.

A two year period of sampling has provided for phase 1 of the project, and at the same time the installation of embayments and weirs has been completed: the location of these structures is indicated in Figure 1. The third phase of the project was begun in April 1986.

3.   PREINTERVENTION CONDITIONS

The sampling programme was initiated in February 1984 with manual sampling of rain, inflows and lake waters; instrumented weirs were installed on sectors IV, VI and VII and the outflow in the summer of 1984. Embayments of the lake were constructed in 1984 (sectors IV and VIII) and in 1985 (sectors V and IX).   Bulk precipitation gauges (British standard) were located at the northwest and southwest (sectors V and IX) and a tipping bucket gauge (MTER) also at sector V.   A weather station (wind speed and direction, temperature, wet-only rain gauge) was installed at the lake outflow.   Manual water sampling is undertaken weekly, lake water sampling monthly, and rain on a daily basis.   At the 4 weirs and the weather station, data are recorded at 15 minute intervals.   This combination of manual and instrumented monitoring provides information on both short term events and sustained conditions.

Rain and water quality conditions for 1985 are shown as mean weighted values in Table 2.   Annual deposition to the catchment is calculated from rain concentrations and volume, and from an estimated dry deposition. Lake effluxes are derived from water concentrations and measured flows (Table 3).

Wet deposition of non sea salt S to the catchment in 1985 is calculated as 1.2 g S/m^2-yr and dry deposition (assuming 5 mg SO_4/m^2 and deposition velocity of 0.5 cm/s) is 0.4 g S/m^2-yr, thus a total of 1.6 g S/m^2-yr.   Deposition of acidity (wet only) is calculated to be 0.038 g H/m^2-yr.

If applied to these data, Henriksen's "acidification index":

$$\text{acidification} = \text{preacidification alk} - \text{present alk} + H^+$$

suggests that sector VII has a value of 82, sector IV(afforested) a value of 37, and the outflow a value of 57.   Thus sector IV is less "acidified" than VI, or the catchment overall.

4.   INTERVENTION OPTIONS

4.1   Several options for intervention are planned at Loch Fleet; others have been considered as a paper exercise.   Those planned or implemented include variants of limestone applications to soils or wetland areas, addition of limestone material to the stream bed, installation of lime dosing equipment on streams, burning vegetation and upper soil horizon, diversion or treatment of forest drainage.   In addition the possible consequences of reduced emissions/deposition can be considered; similarly the possible consequences of forest removal or species replacement.

4.2   The expected results of reduced S emissions may be considered as follows:
    wet deposition (1985)   =   1.2 g S/m^2-yr
    dry deposition (1985)   =   0.4 g S/m^2-yr
    total deposition   =   1.6 g S/m^2-yr

On the basis of similar conditions at the EMEP station some 80 km to the east of Loch Fleet at Eskdalemuir, where rainfall is 1500 mm/yr, and background sulphur level in rain is given as 0.3 mg/l, the total deposition at Loch Fleet is attributed as follows:

    background (0.6/1.6)   = 37%
    CEGB emissions   = 20%
    other UK emissions   = 30%
    continental emissions = 15%

It follows that a possible 65% reduction in CEGB emissions (about 30% of total UK emissions) would result in a reduction of 0.2 g S/m^2-yr, or 12.5%. Even if all attributable S (CEGB, other UK and European emissions) were reduced by 30%, sulphur deposition would be reduced only to 1.2 g S/m^2-yr (44%). It is not known from other sites (e.g the RAIN site in Norway, NIVA, 1985), or Lake Gardsjon in Sweden (Hultberg, 1986) whether the acidity of the resulting drainage water would change, or over what time scale. With UK emissions peaking in 1970 at about 6 m tonnes SO_2/yr (43% greater than in 1984), deposition in 1970 will have been greater than today, say about 2 g S/m^2-yr, assuming no change in background or in continental sources. This recent decline in deposition is concurrent with the reported <u>increased</u> acidity from pH about 5.5 to pH about 4.5. A further, smaller, <u>reduction</u> in deposition seems unlikely to reverse this trend. The argument that the lake requires more than 15 years to respond to a change in deposition is also inconsistent with the predominant (greater than 90%) overland route of drainage flow and the less than 6 months turnover time of the lake water. Nor is there evidence from the calculated fluxes for 1985 that deposited S is retained within the catchment. A further disadvantage of reduced deposition of acid or sulphur is that the level of calcium in the lake water will not be improved - a significant factor for fish survival.

4.3   There is evidence from Scotland (Egglishaw et al., 1986) and Wales (Stoner et al., 1984) that increasing acidity and fishery decline are contemporary with increased afforestation. At Loch Fleet, although only 12% of the overall catchment carries forest, sector IV has 88% forest cover. Water quality draining this sector differs significantly from that from moorland sectors (Table 1) in being somewhat more acid; it has a higher salt concentration, consistent with a greater scavenging of aerosols, but sulphate is lower than that from sector VII. Its drainage is, however, much more variable. Aerosol enriched rain and throughfall collected independently in sector IV (Table 4) over the same period (1985) indicate that all the ions are more concentrated that in bulk precipitation, and that canopy leaching contributes bases to the throughfall. In this N limited system, uptake of ammonium from rain is also thought to add to the acid flux (Leech, pers. comm.). There are, however, doubts about the relationship of this sector's deposition estimate to the standard gauge samples in sectors V(adjacent) and IX. Differences in sampler design and location may explain the differences in rainfall quality evident in Tables 2 and 4. Other explanations may be associated with the complex interaction of deposition with the tree canopy, or to the presence of other vegetation, or possibly to a greater base content of the soils in sector IV.

Notwithstanding these uncertainties, it is possible to consider what would be the effect of deforestation, assuming that sector IV drainage would revert to that of the moorland sectors. Since this sector IV is only 5.6 ha of the total 107 ha, it must contribute only about 5% of the total runoff, and the water quality of the lake as a whole could not be expected to change significantly. Even if a canopy excess of sulphate (say about 130 µeq/1 compared with about 100 µeq/1 in moorland drainage) was transferred to the lake, concentration differences of only about 1.5% would result. Presumably, removal of this "forest effect" would also reduce the yield of calcium to the lake, so that, as with reduced emissions, the water quality for fish would not be improved.

4.4 In contrast, lime applications have a more predictable and immediate effect in improved water quality. Lime applications has been made as follows:

sector IV: 110 tonnes to 5.6 ha as a slurry, 19.6 tonnes/ha
sector VI: 136 tonnes to 6.7 ha uniformly spread, 20 tonnes/ha
sector VII: 105 tonnes to about 10 ha wetland, overall 2.7 tonnes/ha.

Thus a total 351 tonnes have been distributed on 40.7 ha, or 8.6 tonnes/ha of the 3 treated sectors, or 3.3 tonnes/ha for the whole catchment. This treatment has brought the water quality of sector VII to pH about 7, calcium 10 mg/1, aluminium less than 100 µg/1 within 15 days, and has reduced the variations observed during rain events. In sector IV water quality was pH 7.7, calcium 30 mg/1, aluminium 10 µg/1 within 5 days of slurry distribution.

5.   CONCLUSIONS

It seems clear that liming procedures adopted at Loch Fleet are quickly effective and meet the original objectives of the project. It is too early to report the later outcome of treatments, but present results look promising. On the other hand, the possible benefits from emission reduction or deforestation seem uncertain and insignificant for this site.

Changes in land management at Loch Fleet (grazing and burning) or in hydrological management (diverting or treating forest drainage) are worth consideration, since historic records suggest that changes in management occurred over the time of water quality decline. Other procedures may include an increase in lake productivity, and treatment of stream gravels.

REFERENCES
(1)   BATTARBEE, R.W. et al., Report to the Loch Fleet management project, April 1985 (also in Loch Fleet News No. 6)
(2)   BOWN, C.J. and HESLOP, R.E.F. (1979), The soils of the country round Stranraer and Wigtown. Publ. Macaulay Inst. for Soil Research
(3)   Cally Estate Records, Gatehouse-of-Fleet
(4)   EGGLISHAW, H., GARDINER, R., FOSTER, J. (1986), Salmon catch decline and forestry in Scotland, Scott. Geogs. Mag. 102: 57-61
(5)   HULTBERG, H., (1985), Changes in fish populations and water chemistry in L. Gardsjon and neighbouring lakes during the last century, pp. 64-72, Ecol. Bulletins, 37
(6)   LEECH, A. (1986), Report to the Loch Fleet management committee, April 1985 (also in Loch Fleet News No. 6)
(7)   NCC, 1983, Land use history of Cairnsmore of Fleet NNR, publ. Nature Conservancy Council
(8)   NIVA (1985), RAIN project: annual report for 1984, Norwegian Institute for Water Research
(9)   STONER, J.H., GEE, A.S. and WADE, K.R. (1984), The effects of acidification on the ecology of streams in the upper Tywi Catchment in West Wales, Env. Poll (A) 35: 125-157

Table 1: Loch Fleet: General Features

Area of lake:	17 ha
Max. depth:	15 m
Turnover:	~6 mo
Area of catchment:	108 ha
Geology:	Granite, muscovite-biotite
Vegetation	Forest, 12% of total area Calluna, 87% of sample sites Molinia, 75% of sample sites Sphagnum, 15% of sample sites
Rain:	2153 mm yr^{-1} (30 yr mean) pH 4.83 (1985 mean) (3.79-6.00)
Soils:	Peaty gleys, peaty rankers Depth 26-30 cm (median) pH 2.5-3.5 (CaCl$_2$)
Lake biology:	Invertebrates (102 taxa) Fish - b. trout lost 1950's Macrophytes - Juncus, Lobelia
Lake water:	pH 4.0-4.5 Calcium ~1 mgl^{-1} Aluminium 0.1-0.3 mg l^{-1}

Table 2: Loch Fleet: Vol-weighted Mean Concn.
January 1985-December 1985 (µeq l^{-1})

	H$^+$	Na$^+$	Ca^{++}	Mg^{++}	NH$_4^+$	SO$_4^=$	NO$_3^-$	Cl$^-$
RAIN (bulk)	14	59	8	14	21	46 (39*)	11	67
INFLOWS VIII	34	198	56	48		120 (97*)	18	224
IV	42	250	48	51		113 (86*)	14	264
OUTFLOW	36	210	96	66		143 (113*)	11	288

* Sea salt corrected

Table 3:   Loch Fleet: "Annual" Fluxes for Whole Catchment

(128 ha) February 1985–December 1985 keq ha^{-1} yr^{-1}

	H$^+$	Na$^+$	Ca^{++}	Mg^{++}	K$^+$	NH$_4$$^+$	SO$_4$$^=$	NO$_8$$^-$	Cl$^-$
INPUT (bulk)	0.376	1.035	0.138	0.249	0.027	0.394	0.923	0.227	1.185
OUTPUT	0.244	1.465	0.399	0.389	0.008	0.002	0.860	0.132	1.551
BALANCE (%)	−35	+42	+189	+56	−70	−99	−8	−43	+31

Table 4:   Loch Fleet: Rain, Aerosol Enriched Deposition, Throughfall

Vol.-weighted Means 1985

	pH	Na$^+$	Ca^{++}	Mg^{++}	K$^+$	NH$_4$$^+$	SO$_4$$^=$	NO$_3$$^-$	Cl$^-$
Rain (bulk)	4.53	78	20	82	2.6	28	56	19	85
Aerosol enriched	4.40	313	30	57	7.7	44	104	47	336
Throughfall Sitka Spruce	4.15	161	60	41	36	22	146	31	180
Lodgepole	4.16	170	55	41	31	22	131	34	197

(A. Leech: pers. comm.)

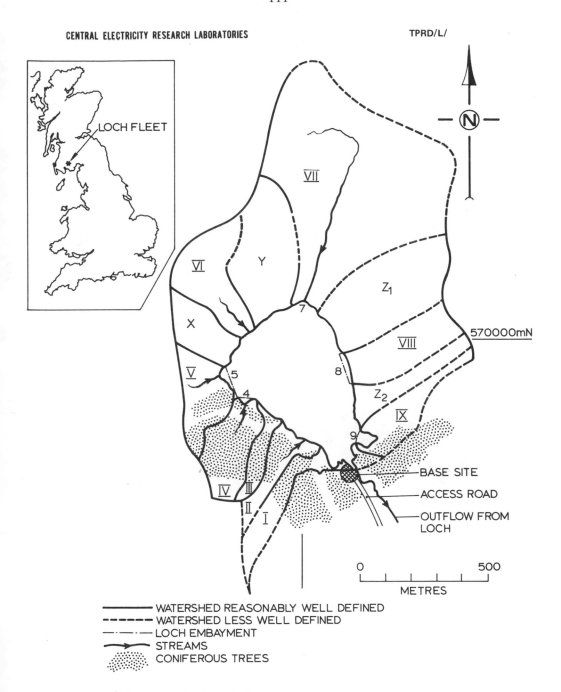

CENTRAL ELECTRICITY RESEARCH LABORATORIES

TPRD/L/

WATERSHED REASONABLY WELL DEFINED
WATERSHED LESS WELL DEFINED
LOCH EMBAYMENT
STREAMS
CONIFEROUS TREES

FIG.1   LOCH FLEET CATCHMENT
SHOWING THE LOCATION OF SUBCATCHMENTS
ASSOCIATED WITH LOCH EMBAYMENTS AS NUMBERED

GDH/HB(9·1·86)RL4·5·3089

SESSION II - THEORETICAL STUDIES AND MODELS

Modelling reversibility of acidification with mathematical models

Modelling stream acidity in U.K. catchments

Testing a soil-oriented charge balance equilibrium model for freshwater acidification

Changes in streamwater chemistry and fishery status following reduced sulphur deposition : tentative predictions based on the "birkenes model"

Simulation of pH, alkalinity and residence time in natural river systems

Evidence of stream acidification in Denmark as caused by acid deposition

MODELLING REVERSIBILITY OF ACIDIFICATION WITH MATHEMATICAL MODELS

B.J. COSBY
Department of Environmental Sciences
University of Virginia

Summary

Quantitative predictions of the effects of reduced acid deposition (the reversibility of acidification) require physically based, process oriented models of catchment soil water and stream water chemistry. A conceptual overview of the soil reactions explicitly or implicity included in many current acidification models is presented. A particular model, MAGIC, is used to examine the patterns and time scales of water quality changes following reduced deposition. The utility of mathematical models in theoretical studies of reversibility is discussed.

1. INTRODUCTION

Recent research has focused attention on certain chemical processes in the soils of catchments as likely keys to the responses of surface water quality to acid deposition.

These processes include:
- anion retention by catchment soils (eg., sulfate adsorption);
- weathering of minerals in catchment soils as a source of base cations (Ca^{2+}, Mg^{2+}, Na^+, K^+) and aluminum (Al^{3+});
- adsorption and exchange of base cations and aluminum by catchment soils;
- alkalinity generation by dissociation of carbonic acid with subsequent exchange of hydrogen ions for base cations.

A critical unanswered question, however, is how quickly and to what extent these processes control surface water quality responses to changes in rates of atmospheric deposition of sulfuric acid. Of particular interest to this workshop is the corollary question of the extent and time scales of reversibility of acidification in systems dominated by these soil processes. Hydrological or in-lake processes may control acidification responses in some systems. Since these factors are discussed by other participants in this workshop, I will confine my remarks to a consideration of the effects of catchment soils as they are currently understood.

The water quality changes we are interested in understanding (and ultimately predicting) occur over several years in natural systems. Direct observation of the dynamics of these changes will, in most cases, require the acquisition of lengthy and costly time series of water quality measurements. In the meantime, we face the problem of integrating and understanding the implications (for whole catchment responses) of results from many individual process level studies. We would like to use existing information to estimate the patterns, time scales and magnitudes of long-term changes in surface water quality in response to actual or assumed changes in the levels of atmospheric sulfur deposition. Physically based, process oriented models (either conceptual or quantitative) can be used for this purpose. The effects that soil processes might have on the time scales and magnitudes of surface water acidification can be examined in a series of modelling exercises.

Models of surface water acidification have been built for a variety of purposes ranging from estimating transient water quality responses for individual storm events to estimating chronic acidification of soils and base flow surface water. The details of these models have been published elsewhere (eg., Christophersen and Wright, 1981; Christophersen et al., 1984; Schnoor et al., 1984; Cosby et al., 1985 a,b; Booty and Kramer, 1984; Goldstein et al., 1984; Bergstrom et al., 1985). In this paper I will: 1) present a conceptual overview of the soil reactions included (explicitly or implicitly) in these models; and 2) use one of the models (MAGIC, Cosby et al., 1985 a,b) to examine the effects these reactions have on reversibility of acidification.

2. CONCEPTUAL BASIS OF SOIL EFFECTS MODELS

The most serious effects of acid deposition on catchment surface water quality are thought to be decreased pH and alkalinity and increased base cation and aluminum concentrations. In two papers, Reuss (1980,1983) proposed a simple system of reactions describing the equilibrium between dissolved and adsorbed ions in the soil-soil water system. Reuss and Johnson (1985) expanded this system of equations to include the effects of carbonic acid resulting from elevated CO_2 partial pressure in soils and demonstrated that large changes in surface water chemistry would be expected as either CO_2 or sulfate concentrations varied in the soil water. The conceptual approach of Reuss and Johnson is attractive in that a wide range of observed catchment responses can be theoretically produced by a rather simple system of soil reactions.

Alkalinity is generated in the soilwater by the formation of bicarbonate from dissolved CO_2 and water:

$$CO_2 + H_2O = H^+ + HCO_3^-. \qquad\qquad 1)$$

The free hydrogen ion produced by this mechanism reacts with an aluminum mineral (eq. gibbsite) in the soil:

$$3H^+ + Al(OH)_3 = AL^{3+} + 3H_2O. \qquad\qquad 2)$$

Generally, the cation exchange sites on the soil matrix have a higher affinity for the trivalent aluminum cation than for di- or monovalent base cations. An exchange of cations between the dissolved and adsorbed phases results: eq.,

$$2Al^{3+} + Ca_3X = Al_2X + 3Ca^{2+}. \qquad\qquad 3)$$

The net results of these reactions is the production of alkalinity in the form of bicarbonates of the base cations (eq., $Ca(HCO_3)_2$). As CO_2 partial pressure increases, the equilibrium reactions above proceed farther to the right hand side in each case, resulting in higher alkalinity. CO_2 partial pressures in soil are commonly much higher than atmospheric CO_2 partial pressure.

If the soil water is removed from contact with the soil matrix and is exposed to the atmosphere (i.e. soil water enters the stream channel), the solution will degas CO_2 due to the lower atmospheric partial pressure of

CO_2. However, the solution is no longer in contact with the soil so cation exchange reactions cannot occur. Changing the CO_2 partial pressure of a bicarbonate buffer solution will result in a change of pH but no net change in alkalinity. The alkalinity of the soil solution is equal to the alkalinity of the stream water.

These processes are illustrated in Figure 1A. The arrows indicate the net mass action effect on the equilibrium processes as the partial pressure of CO_2 in the soil increases. BC^+ represents a base cation. In real systems all four base cations are present and have different affinities for the soil exchange sites. Also, the dissolved trivalent aluminum can complex with dissolved anions (eq., SO_4^{2-} or F^-) or can be hydrated to form dissolved $Al(OH)^{2-}$, $Al(OH)_2^+$, $Al(OH)_3$ and $Al(OH)_4^-$. These complexities are ignored for the moment. As indicated in Figure 1A, the exchange equilibria occur only in the soil solution.

If there are no exchangeable base on the soil matrix the situation is different (Figure 1B). Production of HCO_3^- and H^+ from dissolved CO_2 and dissolution of $Al(OH)_3$ proceed as before. However, there is no longer any possibility of exchange of Al^{3+} for base cations (this situation may occur before all base cations are lost from the soil if the soil affinity for base cations is large). The soil solution in this case consists of HCO_3^- and Al^{3+} ions. The Al^{3+} ion contributes to the acidity of the soil solution. Each Al^{3+} ion is the equivalent of 3 hydrogen ions so that an abbreviated definition of the alkalinity of the soil solution is given by:

$$alk = (HCO_3^-) - (H^+) - 3(Al^{3+}), \qquad 4)$$

where the parentheses represent molar concentrations. The aluminum acidity is exactly balanced by the bicarbonate alkalinity. When the soil water enters the stream, CO_2 degasses consuming one bicarbonate and one hydrogen ion for each molecule of CO_2 lost. As the concentration of hydrogen ions decreases, the solubility of the aluminum solid phase is exceeded and Al^{3+} precipitates as $Al(HO)_3$, releasing H^+ (the reverse of reaction 2). These reactions proceed until a new equilibrium is reached. If the same form of $Al(OH)_3$ precipitates in the stream as was dissolved in the soil, there is no net change in alkalinity. Without exchangeable base cations on the soil the net result of the model processes is zero alkalinity.

Consider now what occurs when an external source of strong acid (atmospheric deposition) is added to the soil. Figure 2A shows the addition of H_2SO_4 to a soil with exchangeable base cations. For the moment, we will assume that the SO_4 concentration is not affected as precipitation enters the soil (i.e., no sulfate adsorption). The mass action as a result of high partial pressures of CO_2 continues as before. The hydrogen ions from atmospheric deposition dissolve additional Al^{3+} which in turn forces the cation exchange reactions to proceed further. However, the cation exchanges are equilibrium reactions. As the base cation concentrations increase, relatively less of the Al^{3+} will be exchanged. The Al^{3+} remaining in solution represents acidity which will reduce the net alkalinity from the situation with no acidic deposition. The amount of additional Al^{3+} that can be exchanged depends on the amount of exchangeable base cations on the soil. Soils with a large amount of exchangeable base cations will respond to acid deposition by neutralizing essentially all of the atmospherically derived H^+. Soils with a small amount of exchangeable base cations will be able to neutralize little of the atmospherically derived acidity. In either case, the effects of acidic deposition are a partial reduction of alkalinity and an increase in base cation concentrations.

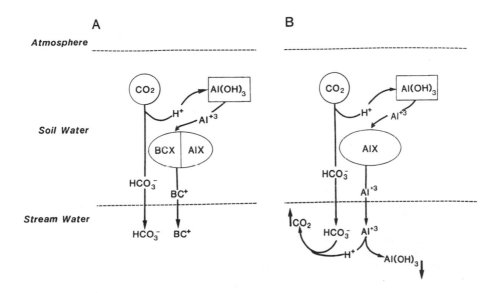

Figure 1. Conceptual soil reactions in undisturbed system: A, exchangeable base cations; B, base cations depleted.

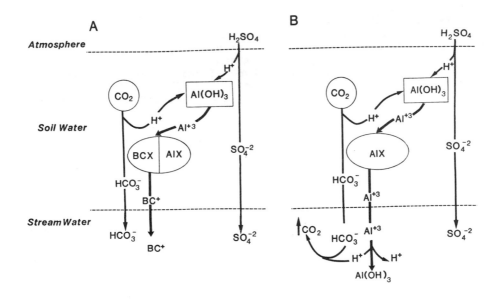

Figure 2. Conceptual soil reactions in system receiving acid deposition: A, exchangeable base cations; B, base cations depleted.

The additional base cations are necessary to balance the negative charges on the mobile anion, SO_4^{2-}, which passes through the system. This is the well known salt effect. As the soil solution degasses on entering the stream, there is again no change in the net alkalinity of the soil solution. The base cation salts of bicarbonate and sulfate remain totally dissociated as the pH rises. The initial effect of adding a strong acid to the system is to increase the ionic strength of the streamwater. Stream alkalinity is partially reduced. The magnitude of the reduction depends on the amount of exchangeable base cations on the soil.

If the H_2SO_4 is added to the soil with no exchangeable base cations (Figure 2B), the acidity of the precipitation is not buffered. As before, the SO_4^{2-} passes through the system. The hydrogen ion deposited from the atmosphere dissolves Al^{3+}. No cation exchange occurs so all of the dissolved Al^{3+} enters the stream. When the solution degasses, that portion of the Al^{3+} produced by the carbonic acid is consumed as $Al(OH)_3$ precipitates. The excess Al^{3+} produced by the atmospheric hydrogen ion is not balanced by an equivalent amount of alkalinity. As the stream pH rises, these excess Al^{3+} ions precipitate, producing free H^+ ions in the stream. The net result is acid streamwater (H^+ and SO_4^{2-}) with a lower pH and higher Al^{3+} concentration than in the undisturbed case.

The discussion so far has been concerned only with the initial mass action shifts in the equilibrium processes. These equilibria are assumed to occur instantaneously. The question arises: what controls the long term response of the catchment streamwater chemistry? Clearly, in the undisturbed case (Figure 1) the situation in a catchment would be expected to shift from that in Figure 1A to that in Figure 1B if there were no long term supply of base cations to replace those lost from the exchange sites. That long term supply must be the weathering of primary minerals in the catchment soils. If the system had been operating long enough to achieve a steady state, the output flux of base cations in the stream would equal the primary weathering input flux. For catchments with a primary mineral source of base cations, the steady state condition will resemble Figure 1A. The degree of base saturation (fraction of soil cation exchange sites occupied by base cations) at steady state would be a function of the primary weathering rate, the cation selectivity of the soil and the hydrological response of the catchment.

If the steady state catchment is suddenly subjected to acid deposition (Figure 2A), the excess base cations produced by the salt effect must be derived from the exchangeable base cations on the soil. This assumes that primary mineral weathering is not increased by the acid deposition. (This assumption seems valid since the net effect of the soil processes is to buffer the soil pH. Changes in soil pH will lag the onset of acid deposition. Unless soil solution pH changes, primary weathering will not be affected). The increased loss of base cations from the catchment will move the system away from the steady state. The base saturation of the soils will decline and the system will move from the situation depicted in Figure 2A to that shown in Figure 2B. If the acid deposition remains constant, the stream base cation concentrations will eventually begin to decline after the initial increase due to the salt effect. When a new steady state is reached, the stream base cation concentrations will have returned to their pre-acidification levels (stream output of base cations equals unchanged weathering input of base cations). The increased mobile anion charge will be balanced by H^+ and Al^{3+} and the stream alkalinity and pH will have declined.

The steady state conditions require that the alkalinity of any catchment be reduced by an amount equal to the acidity of the precipitation when a new steady state is achieved. If deposition acidity exceeds the alkalinity of the catchment, the stream becomes acidic. The crucial questions are: How long will it take to reach the new steady state? What happens to the system during the transition? It is to address these questions that mathematical models of soil processes are developed.

3. A MODELLING EXPERIMENT USING MAGIC: REVERSIBILITY OF ACIDIFICATION.
   (The following discussion is adapted from the published work of Cosby et al., 1985c)

The hypothetical catchment we use in this experiment was designed to be sensitive (low alkalinity prior to onset of acid deposition) yet relatively slow responding (moderate sulfate adsorption capacity). In fact, the values used for the parameters in MAGIC for this exercise are similar to the values used by Cosby et al. (1985 b) for the White Oak Run catchment in Virginia. A square wave (on-off) sulfur deposition sequence is used to drive the model. The deposition square wave is constructed using currently measured atmospheric depositions at White Oak Run for the "on" portion of the sequence and estimated background atmospheric deposition at White Oak Run for the "off" portion. The model is initially in steady state with the background deposition. After 20 years the deposition is assumed to increase in one year to presently observed levels. the acid deposition remains "on" at this level for 120 years. The deposition is then decreased again to background levels and remains "off" for the next 140 years to allow examination of the reversibility of acidification.

The output of the model (Fig. 3) is broken into the stages of acidification and recovery proposed by Galloway et al. (1983) as a convenient means of comparing time scales of water quality changes.

Stage 1. Preacidification stage. This is the steady state prior to the increase in atmospheric deposition of sulfur and is represented by the first twenty years of model output. Recovery of the hypothetical catchment after deposition ceases is assessed by the rate at which the water quality variables return to these preacidification levels.

Stage 2. Undersaturated sulfate adsorption capacity. During this stage there is a lag in the increase of strong acid anion concentrations (SSA in Fig. 3A, principally SO_4^{2-}) as the soils adsorb the atmospherically derived sulfur. As anion concentrations slowly increase, charge balance requires an increase in base cation concentrations (SBC in Fig. 3A) and a decrease in alkalinity (Alk). Given the sulfate adsorption characteristics of the model catchment (which are typical of many natural catchments in the southeastern U.S. (Johnson and Henderson, 1979; Johnson et al., 1980, 1982), forty to fifty years are required before stream sulfate concentrations stop increasing and the catchment achieves a sulfur input-output mass balance. In this hypothetical catchment the alkalinity is depleted (Alk<0) after approximately 20 years even though the catchment soils are still adsorbing sulfur. The rate of supply of base cations from soil cation exchange reactions reaches a peak after approximately 30 years (as indicated by the plateau and downward turn in base cation concentrations). The response time for the initial acidification of the catchment while it is adsorbing sulfate is on the order of two to four decades.

Stage 3. Saturated sulfate adsorption capacity. Concentrations of anions reach a new steady value as the soil sulfate adsorption sites are filled and the dissolved-adsorbed sulfate equilibrium is re-established. Continued flux of high concentrations of anions through the soils strip the soil of base cations. Base cation concentrations decrease toward a new

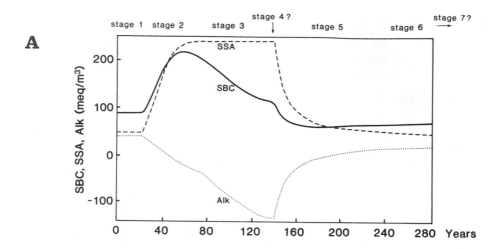

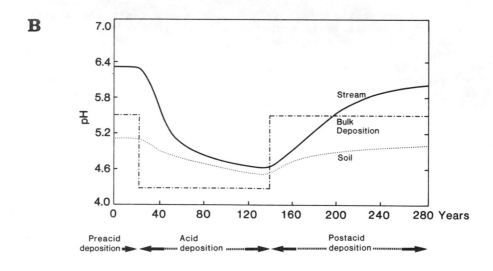

Figure 3. Response of simulated catchment to square wave of deposition: A, changes in sum of base cations (SBC), sum of acid anions (SAA), alkalinity (ALK) and pH of streamwater; B, changes in pH of soil water, streamwater and precipitation.

equilibrium between adsorbed and dissolved phases and alkalinity is further depressed. Even though the catchment reaches a steady value for sulfate in 45 years (stage 2), base cation concentrations and alkalinity continue to decline for an additional 55 years. The response time of continued acidification of the catchment when it is no longer accumulating sulfur from atmospheric deposition is on the order of several (5 or more) decades.

Stage 4. New steady state at the higher level of atmospheric deposition. No further changes in water quality are expected as long as atmospheric deposition remains constant at the higher level. During stage 4, base cation concentrations should return to preacidification levels because at steady state these levels are controlled only by the primary mineral resupply rate. In fact for this model run, the base cation concentrations have not returned to the preacidification levels before the deposition is turned off. Stage 4 is not achieved in this example. However, the rates of change of base cation concentrations and alkalinity decrease markedly near the end of stage 3 and appear to be approaching asymptotic values by the 140th year of the simulation. The acidification responses of this hypothetical catchment to an approximately tenfold increase in sulfur deposition required on the order of one century to complete.

Stage 5. Supersaturated sulfate adsorption capacity. This stage begins when the deposition is decreased to the initial low level. The rate of decline of streamwater anion concentrations will depend on conditions in the soil because the sulfate adsorbed by the soil must be flushed from the system. As the anion concentrations decrease, the base cation concentrations also decrease and alkalinity begins to increase as fewer hydrogen and aluminum ions are needed to balance the strong acid anions. The acidification and recovery curves produced by MAGIC are notably asymmetric. The recoveries estimated by the model are initially rapid but become progressively slower. There are no lags in the initial water quality responses to decreased deposition. During stage 5, base cation concentrations decrease below pre-acidification levels and begin to increase again before the accumulated sulfate has completely desorbed from the soil. The model shows a return of strong acid anion concentrations to pre-acidification levels by the 280th year. Stage 5 lasts for approximately 100 years. Desorption of the SO_4^{2-} accumulated in the soils takes approximately twice as long as the initial adsorption.

Stage 6. Recovery of soil base saturation. After the anions have returned to their preacidification levels, there is a further lag in the recovery of base cation concentrations and alkalinity. Part of the re-supply of base cations from mineral weathering is being used to replace base cations on the soil exchange sites. Until the soil base saturation has returned to preacidification levels, recovery of stream base cation concentrations and alkalinity will not be complete. Although the rates of change of all variables in the model are very modest by the end of this experiment with MAGIC, the values for some of these variables are still far from their pre-acidification levels. The recovery process appears to take approximately twice as long as the initial acidification. The order of magnitude of recovery time for this hypothetical catchment is 1.5 to 2 centuries.

Stage 7. Return to preacidification steady state. This stage is only achieved after an additional 80 years of simulation beyond that shown in Figures 3 and 4. (A subjective judgment must be made concerning approaches to a new steady state. Such approaches are asymptotic and thus Stages 4 and 7 can only be defined by reference to some arbitrary percentage differences from the asymptote).

Given that the modelling experiment has indicated that acidification and recovery response times for this hypothetical catchment are on the order of decades, it is instructive to examine the time scales of the corresponding pH changes (Fig. 3B). The responses of both soil water and streamwater pH are also asymmetric. However, the magnitudes of the changes in soil water pH are much smaller than the corresponding changes in surface water pH. The response times of soil water and streamwater pH are also on the order of decades. As with the other water quality variables, the recovery of pH takes much longer than the initial depression.

4. PROCESSES CONTROLLING REVERSIBILITY

The predicted catchment dynamics for acidification and recovery result from different soil processes. Acidification is controlled by the sulfate adsorption characteristics of the soil and the affinity of the soil for adsorbed base cations. The recovery is controlled by the base cation resupply rate from mineral weathering processes.

The dynamics of sulfate adsorption are perhaps best appreciated by considering the difference between the sum of the strong acid anion concentrations in streamwater (SSA) and the sum of strong acid anion concentrations expected in the streamwater if the system were in steady state with atmospheric inputs (SAE) (Fig. 4A). The small difference between SAE and SSA during the initial steady state is due to the net biological uptake of NO_3^- included in the model. the asymmetry of overall catchment response is being driven by the asymmetry of the sulfate adsorption/desorption dynamics, i.e., by the inherent nonlinearity in the sulfate adsorption process (eg., see Cosby et al., 1986).

Base cation dynamics are influenced by atmospheric inputs, primary weathering inputs and by the aggregate effects of the various soil equilibrium/exchange processes in the model. The total base cation concentration in streamwater (SBC) exceeds the concentration that would be expected (BCE) if the catchment were in steady state and if atmospheric inputs were the only source of base cations (Fig. 4B). The difference between SBC and BCE during the initial steady state (stage 1) is due to the primary mineral weathering inputs of base cations. The initial increase in SBC following the onset of deposition is the salt effect in response to increased sulfate concentrations in the soil water. As the salt effect leaches base cations from the soil (stages 2 and 3), the base saturation of the soil declines (Fig. 4C). At the beginning of stage 5, as sulfate concentrations begin to decline, fewer base cations are exported by the stream and the products of mineral weathering process begin to replenish the exchangeable base cations on the soil. The prolonged recovery of the catchment (relative to the response to increased deposition) results form the slow replenishing of the soil cation exchange sites (Fig. 4C).

Taken collectively all of the responses described above provide a consistent picture of catchment response to atmospheric deposition of sulfate. The mobile anion, sulfate, is either adsorbed, balanced by base cations leached from the soil, or balanced by dissolved aluminum. Sulfate that is adsorbed cannot contribute to the acidification of freshwaters. The amount of sulfate adsorbed is soil dependent. Furthermore, the capacity of soils to adsorb sulfate decreases as sulfate loading continues. The "excess" sulfate that is not adsorbed may be balanced by base cations leached from soil exchange sites. Soil with a large amount of exchangeable base cations will respond to acid deposition by neutralizing essentially all of the atmospherically derived acidity. Soils with a small amount of exchangeable base cations will be able to neutralize little of the atmospherically derived acidity. Thus, we would predict that soils with

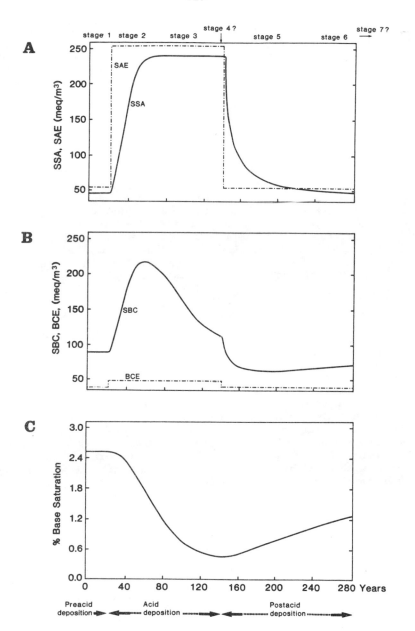

Figure 4. Response of simulated catchment to square wave of deposition: A, simulated (SAA) and atmospheric steady state (SAE), concentrations of strong acid anions; B, simulated (SBC) and atmospheric steady state (SBE) sum of base cation concentrations; C, soil base saturation changes.

low exchange capacity and little sulfate adsorption capacity would characterize catchments of streams that have already been acidified by atmospheric sulfate deposition and that these streams would have response times on the order of a year or less. On the other hand, we would predict that soils with very high cation exchange capacity (eg., most agricultural soils) would characterize catchments with streams that, for intents and purposes, would be immune to the acidification effects. The response time in these cases would be very large (>> 100 yr).

5. DISCUSSION

Are soil process models valid representations of reality? Unfortunately, no entirely satisfactory answer to this question can be given at this time because research on the effects of acid rainfall is beset by two severe problems: (1) the paucity of data and the expense (in terms of time and money) of obtaining additional data and (2) the time scale of the responses in which we are interested. Models of acid deposition effects, be they empirical or mechanistic, simple, or complex, are all developed with the implied goal of estimating catchment responses to changes in deposition rates at some point years or decades in the future. Strict verification of these models would require that we wait years or decades to determine whether the model estimates match the observed catchment responses. Alternately, the models could be verified by using them to reconstruct past catchment water chemistry and comparing that reconstruction with historical observations of water quality in the catchment. This approach is thwarted by the fact that few (if any) records of catchment water quality extend far enough into the past to provide rigorous tests of the models.

Obviously, we are not presently in a position to conclude that our projections derived from models are correct in a quantitative sense. What we can conclude is that models based on processes that are thought to be important in determining the chemical response of catchments to acid deposition (using parameter values that are within the ranges appropriate for natural soils) produce plausible results. As far as present day observations of catchment chemistry are concerned, the models are consistent with available measurements. The question of whether the long-term responses estimated by the models are accurate projections of the responses of the real system must remain unanswered.

Even though quantitative models of catchment response cannot be strictly verified for their ultimate use of long-term estimation, they can be used as heuristic tools to stimulate and inform the synthesis of current research results. ("Are the implied long-term effects of the processes being studied consistent with our conceptual understanding of long-term catchment responses?") Models can be used to bridge the gap between the findings of detailed process level studies and the broader concerns of resource managers and policy makers. ("If the processes being studied are important in determining catchment water quality, what might be the effect of a certain decrease in deposition rates over a particular time?") Models can also be used to feed back into the planning of future research. ("If these processes are indeed controlling catchment response, what kinds and quantities of data may be needed and what experimental design may be required to test our assumptions about the processes?") Such speculative simulation modeling may prove to be the greatest utility of models of whole catchment responses to acidic deposition.

REFERENCES

(1) BERGSTROM, S., B. CARLSSON and G. SANDBERG, Integrated modelling of runoff, alkalinity and pH on a daily basis, Nordic Hydrology, 16, 89–104, 1985.

(2) BOOTY, W.G. and J.R. KRAMER, Sensitivity analysis of a watershed acidification model, Philos. Trans. R. Soc. London, Ser. B, 305, 442–449, 1984.

(3) CHRISTOPHERSEN, N. and R.F. WRIGHT, A model for streamwater chemistry at Birkenes, Norway, Water Resour. Res., 18, 977–996, 1981.

(4) CHRISTOPHERSEN, N., S. RUSTAD, and H.M. SEIP, Modelling streamwater chemistry with snowmelt, Philos. Trans. R. Soc. London, Ser. B, 304, 427–439, 1984.

(5) COSBY, B.J., R.F. WRIGHT, G.M. HORNBERGER, and J.N. GALLOWAY, Modelling the effects of acid deposition: Assessment of a lumped-parameter model of soil water and streamwater chemistry, Water Resour. Res., 21, 51–63, 1985a.

(6) COSBY, B.J., R. F. WRIGHT, G.M. HORNBERGER and J.N. GALLOWAY, Modelling the effects of acid deposition: estimation of long-term water quality responses in a small forested catchment, Wat. Resour. Res., 21, 1591–1601, 1985b.

(7) COSBY, B.J., G.M. HORNBERGER, J.N. GALLOWAY and R.F. WRIGHT, Time scales of catchment acidification, a quantitative model for estimating freshwater acidification, Environ. Sci. Technol., 19, 1144–1149, 1985c.

(8) COSBY, B.J., G.M. HORNBERGER, R.F. WRIGHT and J.N. GALLOWAY, Modeling the effects of acid deposition: control of long-term sulfate dynamics by soil sulfate adsorption, Wat. Resour. Res. (in press), 1986.

(9) GALLOWAY, J.N., S.A. NORTON and M.R. CHURCH, Freshwater acidification from atmospheric deposition of sulfuric acid: a conceptual model. Environ. Sci. Technol., 17, 541A–545A, 1983.

(10) GOLDSTEIN, R.A., S.A. GHERINI, C.W. CHEN, L. MAK, and R.J.M. HUDSON, Integrated acidification study (ILWAS): A mechanistic ecosystem analysis, Philos. Trans. R. Soc. London, Ser. B, 305, 409–425, 1984.

(11) JOHNSON, D.W., and G.S. HENDERSON, Sulfate adsorption and sulfur fractions in a highly weathered soil under a mixed deciduous forest, Soil Sci., 138, 34–40, 1979.

(12) JOHNSON, D.W., G.S. HENDERSON, D.D. HUFF, S.E. LINDBERG, D.D. RICHTER, S. SHRINER and J. TURNER, Cycing of organic and inorganic sulphur in a chestnut oak forest, Oecologia, 54, 1141–148, 1982.

(13) REUSS, J.O. Simulations of soil nutrient losses resulting from rainfall acidity, Ecol. Model, 11, 15–38, 1980.

(14) REUSS, J.O., Implications of the Ca-Al exchange system for the effect of acid precipitation on soils, J. Environ. Qual., 12, 591–595, 1983.

(15) REUSS, J.O., and D.W. JOHNSON, Effect of soil processes on the acidification of water by acid deposition, J. Environ. Qual., 14, 26–31, 1985.

(16) SCHNOOR,J.L., W.D. PALMER, Jr. and G.E. GLASS. Modeling impacts of acid precipitation for northeastern Minnesota, edited by J.L. Schnoor, pp. 155–173, in Modeling of Total Acid Precipitation Impact, Butterworth, Boston, Mass., 1984.

MODELLING STREAM ACIDITY IN U.K. CATCHMENTS

by

P.G. Whitehead
C. Neal and
R. Neale

Institute of Hydrology
Wallingford
U.K.

SUMMARY

As part of the joint British-Scandinavian Surface Waters Acidification Programme, the Institute of Hydrology is establishing catchment studies in Scotland and Wales. Data from these catchment studies are being used to develop a range of models for investigating short term and long term changes in catchment acidity. Information on the modelling techniques available at IH is presented together with applications of the models to catchments in Scotland and Wales. They provide examples of how both acidic deposition and conifer afforestation can increase streamwater acidity. Long term trends are predicted. A 50% reduction in current industrial emissions is required to prevent further increase in stream acidity in the Southern Uplands of Scotland.

1. INTRODUCTION

Catchment studies investigating the acidic behaviour of upland streams are expensive, time consuming and difficult to establish due to the complexity of hydrological, chemical and biological interactions. Nevertheless many catchment studies have been and are being established to evaluate short term and long term fluctuations in stream water chemistry. For example as part of the joint Scandinavian – British Surface Water Acidification Programme (Mason and Seip, 1985) major studies are being established in the UK and Scandinavia. Other organisations such as the Welsh Water Authority (Llyn Brianne Study; Stoner et al, 1984) the Solway River Purificatin Board (Loch Dee study; Burns et al, 1982) and the Freshwater Fisheries Laboratory (Loch Ard Study; Harriman et al, 1981) have also established catchment studies following mounting concern over the loss of fisheries in Scotland and Wales and the possible detrimental effects of stream acidity on water resources. Several researchers involved in these studies (Harriman et al, 1981, Gee and Stoner, 1984) have reported elevated acidity and aluminium levels in upland streams draining afforested (conifer) catchments in the UK. Moreover in many of these areas and particularly forested catchments fisheries have deteriorated and restocking programmes have been unsuccessful.

It is with these problems in mind that IH has established a catchment study in Wales at Plynlimon (see Neal et al, 1985). IH is also establishing a catchment study in the Cairngorm region of Scotland in collaboration with DAFS (Department of Agriculture and Fisheries for Scotland), the Macaulay Institute of Soil Science and Imperial College,

Department of Civil Engineering. IH is responsible for providing stream gauging, rainfall stations, a weather station, snow surveys, sampling and continuous water quality monitoring. IH is also responsible for the subsequent data management, analysis and interpretation. DAFS is responsible for all chemical and biological analysis, with the exception of snowmelt chemistry, which would be undertaken by IH. The Macaulay Institute is responsible for soil-surveys and soil-water chemistry and Imperial College are establishing plot studies.

The hydrological and chemical data collected from the catchment studies forms the basis of a comprehensive modelling research programme by IH.

Recently there has been considerable use of mathematical models to describe the dominant interactions and processes operating in catchments and to simulate catchment behaviour. Steady state models have been used prescriptively to demonstrate the long term consequences of changes in the industrial emissions of SO_2 (Cosby et al, 1985a, b, Kamari et al 1984). Correspondingly, dynamic models have been successfully applied descriptively to several catchments (Christopherson et al, 1982, 1984). For example, Christophersen et al. have developed a simple conceptual model that reproduces major trends in chemical and hydrological behaviour in Norwegian catchments. This model has been successfully extended (Seip et al, 1985) and applied descriptively to the Harp Lake catchment in Canada. The model has also been applied to two forested catchments in Sweden (Grip et al, 1985).

A wide range of mathematical modelling techniques are available at IH for analysing catchment data. These include CAPTAIN (Computer Aided Package for Time Series Analysis and the Identification of Noisy Systems; Venn and Day, 1977, Whitehead et al., 1986), MIV (Multivariable time series model, Young and Whitehead, 1977), the BIRKENES model (Christophersen et al., 1982), MAGIC (Cosby et al., 1985a), EKF (Extended Kalman Filter, Beck and Young, 1976) TOPMODEL (Beven, 1982) and IHDM (Institute of Hydrology Distributed Model, Morris, 1980).

In this paper several of these techniques are described and applied to investigate short term catchment responses or 'events' and long term acidity of soil and stream waters.

2. TIME SERIES OR 'EVENT' TYPE MODELS

Time series models are suitable where the overall input-putput behaviour is of prime importance and where internal mechanisms are particularly complex. It is assumed that a 'law of large systems' applies (Young, 1978) whereby the combination of all the complex non-linear and distributed elements gives rise to an aggregated system behaviour that is relatively simple in dynamic terms.

Application to Loch Dee

Loch Dee has a remote setting in the Galloway Hills in South West Scotland. The catchment is made up of three sub-basins: Dargall Lane to the west, White Laggan Burn with its tributary the Black Laggan towards the south, and Green Burn entering from the south east. The outflow at the north east end of the loch is the source of the River Dee and up to

this itself occupies 1.0 km^2. Catchment altitudes range from 225 m on the loch shore to 716 m on Lamachan at the head of the Dargall Lane. Nearly two thirds of the catchment lies above 305 m (1000 ft). Geologically the area comprises two distinct rock types: Ordovician greywackes/shales and granites of Old Red Sandstone Age (Burns et al., 1982).

In Loch Dee an extensive record of hydrological water quality data has been collected over a five year period (Burns et al., 1982, Langan, 1985). Analysis has been restricted initially to a time series model relating flow to hydrogen ion concentration for the White Laggan sub-catchment. The White Laggan is subject to episodic acidification, primarily attributed to atmospheric inputs (Langan, 1985).

The model fitted is an autoregressive moving average type of the form

$$x_t = -\delta_1 x_{t-1} + \omega_0 u_t$$

where x_t is the hydrogen ion concentration (μeql^{-1}) and u_t is the flow (m^{3}sec^{-1}) in the stream at time t.

The parameters δ_1 and ω_0 were estimated using a time series algorithm applied to 200 **hourly** observations of pH and flow. The parameters were estimated to be;

$\delta_1 = -0.680$ (standard error 0.012)
$\omega_0 = 0.659$ (standard error 0.022)

and Figure 1 shows the simulated hydrogen concentration against the observed concentration. A remarkably good fit to the data is obtained with 93% of the variance explained and suggests that H$^+$ion and flow are closely related. However a true test of the model would be to use an additional data set; Langan (1985) has applied the approach to all three subcatchments of Loch Dee and found that equally good models have been obtained for a wide range of storm events. In the case of the White Laggan a mean response time (T) of 2.6 hours is obtained, reflecting the fast response time between output flow and hydrogen ion concentrations.

Further applications of the time series techniques to data from Welsh and Norwegian Catchments are given by Whitehead et al (1986).

3. APPLICATIONS OF THE 'BIRKENES' MODELS

A second class of models have been applied to data from the Loch Dee study. These include the 'Birkenes' model developed by Christophersen et al(1982, 84).

The model comprises of a simple two-reservoir hydrological model operating on a daily timestep upon which has been superimposed the importantchemical processes that control the acidification of catchments. Inputs to the model are precipitation, mean daily temperature, mean daily soil temperature and sulphate deposition rates. Figure 2 shows the principal hydrological and chemical processes operating.

The model outputs daily concentrations of Hydrogen ion, Aluminium,

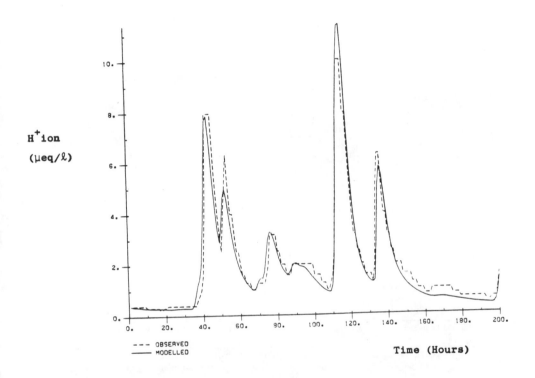

Figure 1 : Simulated and Observed H$^+$ ion in the White Laggan, Loch Dee, Scotland, based on the Flow model

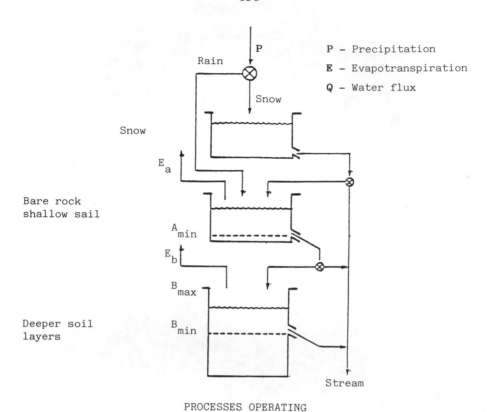

PROCESSES OPERATING

	Shallow soil reservoir	Deeper soil reservoir
H_2O	Precipitation, evapotranspiration, infiltration to lower reservoir, discharge to stream	Infiltration, evapotranspiration, discharge to stream, piston flow
SO_4	Wet + dry deposition, adsorption/desorption, mineralization	Adsorption/desorption, reduction
Ca + Mg	Ion-exchange	Release by weathering, adsorption/desorption
H^+	Ion exchange and equilibrium with gibbsite	Consumption by weathering, adsorption/desorption, equilibrium with gibbsite/
Al^{+3}	Equilibrium with gibbsite	Equilibrium with gibbsite, adsorption/desorption
HCO_3^-		Equilibrium with a seasonal varying CO_2 pressure

Figure 2 : Hydrological model used for Harp Lake Catchment and main processes operating

Sulphate, Calcium + Magnesium (M^{2+}) and Bicarbonate in the stream along with predicted flow.

Numerous modifications and additions over the years have resulted in the program structure becoming somewhat confused. Therefore the model has been extensively re-written at IH so as to greatly increase computational efficiency and improve the readability of the FORTRAN 77 source code and hence simplify the task of making modifications in the future.

The ' Birkenes' model has also been used to assess the sensitivity of stream acidity to hydrological parameters and changes in baseflow.

Flow movement between the soil and groundwater compartments is restricted by a "percolation" equation as follows;

$$A_{SIG} = P - (B - B_{min})/B_{max} \qquad \text{for } B \geqslant B_{min}$$

$$\text{and } A_{SIG} = 1.33P - 0.33P(B/B_{min})^3 \qquad \text{for } B < B_{min}$$

where B refers to the groundwater compartment water level and B_{min} and B_{max} refer to minimum and maximum water levels respectively minimum and maximum water levels respectively (see Figure 2). The parameter, P can be considered as a percolation parameter so that increasing P increases the fraction of flow, A_{SIG}, routed to the lower reservoir. This leads to an increase in the baseflow contribution to the stream. The model also includes a piston flow component to describe the hydraulic movement of water out of the groundwater compartment.

The changes in stream water concentrations for H^+ ion and Al^{3+} to changes in the baseflow contribution to streamwater is highly non linear. This is illustrated in Figures 3 and 4 which show H^+ and Al^{3+} maximums and means over a range of baseflow conditions for both the long and short term data sets: all concentration values fall as the baseflow increases; the rate of decrease varying from one variable to another. In general increases in baseflow results in significant reductions in H^+ ion and Al concentrations. Studies by Seip and Rustad (1983) show a similar non linear behaviour when upper and lower soil horizon waters are mixed.

Further results are given by Whitehead et al (1986) and confirm the sensitivity of the model to parameter, and hence baseflow, changes.

4. APPLICATION OF MAGIC TO LOCH DEE

MAGIC (Model of Acidification of Groundwater In Catchments; Cosby et al, 1985) is explicitly designed to perform long term simulations of changesin soilwater and streamwater chemistry in response to changes in acidic deposition. The processes on which the model is based are:

-   anion retention by catchment soils (e.g. sulphate adsorption):
-   adsorption and exchange of base cations and aluminium by soils;
-   alkalinity generation by dissociation of carbonic acid (at high CO_2 partial pressures in the soil) with subsequent exchange of hydrogen ions for base cations;

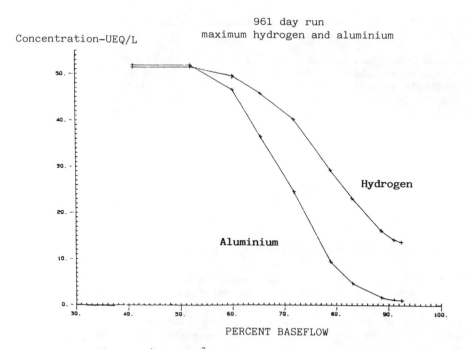

Figure 3 : Maximum H^+ and $A\ell^{3+}$ concentrations in the stream showing variations over a range of baseflow conditions (Three Year Simulation
1977–1980)

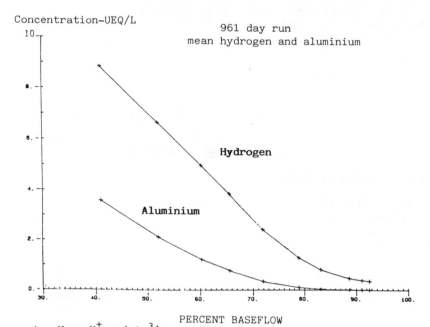

Figure 4 : Mean H^+ and $A\ell^{3+}$ concentrations in the stream showing variation
over a range of baseflow conditions (Three Year Simulation
1977–1980)

- weathering of minerals in the soil to provide a source of base cations;
- control of Al^{3+} concentrations by an assumed equilibrium with a solid phase of $Al(OH)_3$.

A sequence of atmospheric deposition and mineral weathering is assumed for MAGIC. Current deposition levels of base cations, sulphate, nitrate and chloride are needed along with some estimate of how these levels had varied historically. Historical deposition variations may be scaled to emissions records or may be taken from other modelling studies of atmospheric transport into a region. Weathering estimates for base cations are extremely difficult to obtain. Nonetheless, it is the weathering process that controls the long term response and recovery of catchments to acidic deposition and some estimate is required.

The MAGIC program has been applied to the Dargall Lane sub-catchment in Loch Dee and a detailed description of the application is given by Cosby et al (1986).

Several chemical, biological and hydrological processes control stream water chemistry. These processes are often interative and not easily identifiable from field observation. Modelling allows separation of the different factors and the establishment of their relative importance quantitatively. Here the factors considered are afforestation, dry and occult deposition, variations in acidic oxide loading and deforestation.

LONG TERM ACIDIFICATION TRENDS FOR DARGALL LANE

Figure 5 shows a simulation of long term acidity for the Dargall Lane catchment. The sulphate deposition history is shown in Figure 5a and this 'drives' the MAGIC model. The historical simulation of pH shown in Figure 5b is similar to the values obtained from the diatom records of lochs in the region in that a significant decrease in pH from 1900 onwards is inferred (Batterbee et al, 1985, Flower et al, 1983). The steeper decline from 1950 to 1970 follows from the increased emission levels during this period. The model can also be used to predict future stream water acidity given different future deposition levels. For Dargall Lane stream acidity trends are investigated assuming two scenarios for future deposition. Firstly assuming deposition rates are maintained in the future at 1984 levels, the model indicates that annual average stream pH is likely to continue to decline below presently measured values. Second, assuming deposition rates are reduced by 50% from 1984 levels (between 1985 and 2000) the results indicate that current stream water acidity will be maintained (Figure 5b). Further details of the application of this model are given elsewhere (Cosby et al, 1986). Note an increase in stream water pH about 1980; this follows a significant drop in sulphur emissions during the 1970s. Note also that an earlier decline in streamwater acidity is predicted if there had been no reductions in emissions since 1970.

AFFORESTATION

Afforested systems are more complex to model than grassland systems because the introduction of the forest pertubs a grassland ecosystem which in itself is difficult to model. The effects of the forest root system, leaf litter layer and drainage ditches will change the hydrological

5 a

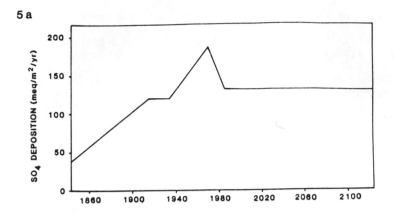

5 b

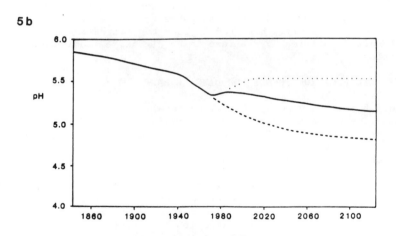

Figure 5 a : Sulphate deposition history used as input for the MAGIC reconstruction of pH in the Dargall Lane moorland catchment

b : Simulation of the pH of streamwater in the Dargall Lane moorland catchment assuming three sulphate deposition scenarios

‾‾‾‾ Historical levels to 1984 and constant 1984 levels thereafter (see Figure 2a)

.... Historical levels to 1984 and 1984 levels reduced by 50 % by the year 2000 and constant thereafter

---- Historical levels to 1970 and constant 1970 levels thereafter

pathways; this will control the nature and extent of the chemical reactions in the soil and bedrock. Further, the additional filtering effect of the tree on the atmosphere will enhance occult/particle deposition and evapotranspiration will increase the concentration ofdissolved components entering the stream. The magnitude of these different effects varies considerably; for example evapotranspiration from forests in the British Uplands is typically of the order of 30% of the precipitatation which is almost twice the figure for grassland. This will have the consequence that the total anion concentrations within the stream and soil waters increase by 14% following afforestation. The forest will also increase anion and cation loading due to the enhanced filtering effect of the trees on air and occult sources. [Several forest catchment studies have shown that Chloride concentrations are higher in streamwaters than in the corresponding rainfall or grassland streams. Such a difference results from dry and occult deposition assuming that chloride is derived from a maritime source and not from leaching from the catchment bedrock. Such increases are typically of the order of 30%]. The filtering effects will apply both to marine and pollutant aerosol components. Altering the hydrological pathways can also have a major effect on stream water quality since the forest tends to increase surface runoff thereby flushing/ displacing highly acidic water from the surface layers; the soil zone acts as a proton and aluminium source whilst the bedrock, if silicate or carbonate bearing, provides proton consumption by weathering reactions.To illustrate the effects of afforestation simply in terms of increased concentrations from both enhanced dry deposition and evapotranspiration, the MAGIC model has been applied to the Dargall Lane catchment assuming that a forest is developed over the next 40 years. It should also be noted that, here, no allowance has been made for the effects of cation and anion uptake by the trees during their development; the incorporation of base cations into the biomass would result in an enhanced acidification effect during this period.

Of critical importance is the relative and absolute contribution of marine and pollutant inputs from dry and occult deposition. Figure 6 shows the effects of increasing evapotranspiration from 16 to 30% over the forest growth period with varying levels of marine, pollutant and marine + pollutant inputs. Increasing either marine or pollutant components leads to enhanced stream water acidity, the greatest effects being observed when both components are present; the effect of simply increasing evapotranspiration from 16 to 30% has a similar effect but the changes are much smaller. The important features of these results are the enhanced and acidic oxide inputs from increased scavenging by the trees result in a marked reduction in pH levels and that there is an additive effect when both processes are combined. These reductions are much greater than the effect of evapotranspiration.

ATMOSPHERIC ACIDIC OXIDE INPUTS

An important factor in determining stream acidity in the upland UK is the level of acidic oxide deposition; rates of deposition (non marine wet deposition and dry deposition)[14-15] can vary from 0.5 to over 6 g S $m^{-2}yr^{-1}$ and from 0.1 to over 0.5 g N $m^{-2}y^{-1}$. Figure 7 shows the effects of such variations for both moorland and forested catchments; the highest levels correspond to areas with high atmospheric acidic oxide rates (3 times the historic and 1984 deposition levels observed in the Southern Uplands of Scotland). With increasing atmospheric acidic oxide pollution,

INCREASING DRY/OCCULT DEPOSITION

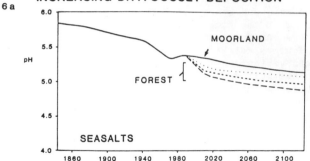

6 a

Figure 6 a : Simulation of the pH of the streamwater from the Dargall Lane catchment
comparing the moorland catchment response assuming Figure 2a deposition
rates (——), the effect of 14 % additional evaporation following afforestation (...), the effect of 14 % additional evaporation plus 15 %
additional input of natural seasalts following afforestation in 1985
(–––), and the effect of 14 % additional evaporation plus 30 % additional
input of natural seasalts following afforestation in 1985 (– – –)

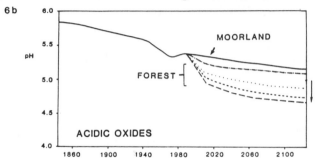

6 b

Figure 6 b : Simulation of the pH of streamwater from the Dargall Lane catchment
comparing the moorland response (——) to the forested catchment response
assuming increased evaporation (–·–·–) with different levels of pollutant
scavenging (..., 20 % additional sulphate, ––––, 40 % additional
sulphate, – – – , 60 % additional sulphate)

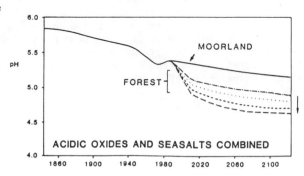

6 c

Figure 6 c : Simulation of the pH of streamwater from the Dargall Lane catchment
showing the moorland response (——) and the combined effects on the
forested catchment of increased evapotranspiration, increased scavenging
of natural sea salts, and various levels of increased scavenging of
pollutants inputs (–·–·–· , zero additional pollutant scavenging, ..., 20
% pollutant scavenging, –––, 40 % pollutant scavenging, – – – , 60 %
pollutant scavenging)

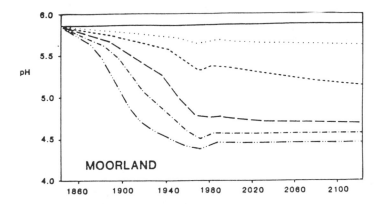

Figure 7 a : Simulation of the pH of streamwater from the Dargall Lane
moorland catchment assuming sulphate deposition patterns
modified by various factors to reproduce a range of loading
conditions (i.e. from pristine to heavy pollution)

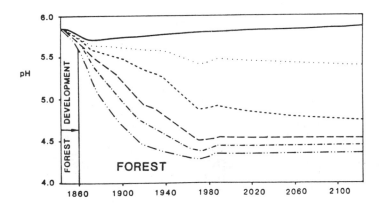

Figure 7 b : Simulation of the pH streamwater from the 'forested' Dargall
Lane catchment assuming afforestation from 1844 onwards and
sulphate deposition patterns (see Figure 5 a) multiplied by
various factors to reproduce a range of loading conditions
from pristine to heavy pollution

_____	Background rates	(pristine conditions)
.....	0.5 x Figure 5 a deposition concentrations	(low pollution)
-----	1 x " " " "	intermediate
_ _ _	1.5 x " " " "	pollution
-.-.-	2 x " " " "	heavy
_.._..	3 x " " " "	pollution

the decline in stream pH is accelerated, the changes occur much earlier, and the final pH of the stream water is lower.

DEFORESTATION

Whilst afforestation increases stream acidity, as shown both by the model predictions and field evidence, then deforestation will result in a reduction in stream water acidity. Figure 8 shows the effects of deforestation from the present time for a range of acidic input loadings. The result shows that while there is a short term improvement in stream acidity, the long term acidification trend is maintained. It is interesting to note that the recovery following deforestation at the intermediate deposition levels is greater than that at the higher levels. This is because base saturation has not been completely depleted, and the reduced deposition following deforestation can be buffered by the available cations. Under the higher deposition levels base saturation is reduced to very low levels making recovery much less significant. Note that afforestation following tree harvesting will negate the improvement in stream water acidity.

IMPLICATIONS

The modelling enables assessment of the relative effects of atmospheric acidic oxide pollution and conifer afforestation, as well as highlighting some of the topics that need further consideration. For example, the long term trends in stream water acidification for the grassland catchment suggest that for at least part of the upland UK, acidic oxide pollutant inputs are the dominant source of increased stream water acidity. The model predictions are similar to observations of stream acidity found in Southern Scandinavia and add weight to the conclusion that such pollutant inputs are also a major source of stream acidification in those countries as well. How important this acidification process is on a regional basis in the upland UK cannot be gauged immediately because many unresolved factors remain, as mentioned above. However, much of the British uplands have soils which are susceptible to acidic inputs; it is therefore reasonable to assume the results of this present modelling exercise are widely applicable. If the above results are representative of sensitive upland areas then reductions in present acidic emissions of the order of 50% are required to prevent further increase in stream acidity moorlands; afforested catchments require greater reductions. The study points to the need for further regional analysis of soil and stream water chemistry, as well as a better understanding of hydrogeochemical processes operating within catchments. Further, the study provides an example of the need to establish the extent of scavenging of aerosols onto plant surfaces, and more generally on the benefits of multidisciplinary catchment studies. Finally, the detrimental effect on stream water quality caused by conifer afforestation in uplands subject to acidic deposition is irrefutable. While there is uncertainty regarding the nature and the extent of the hydrogeochemical processes operative there is a need to change existing forestry practices which are of immediate pragmatic concern.

5. CONCLUSIONS

The model techniques applied at IH have proved to be particularly useful yielding information on the catchment responses, processes and possible future behaviour.

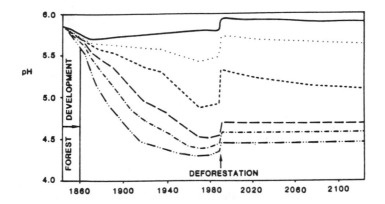

Figure 8 : Simulation of the pH of streamwater for the Dargall Lane
catchment assuming afforestation from 1844 and deforestation in
1990.
Deposition patterns as described in Figure 7

On the hydrological side time series techniques, lumped and
distributed hydrological models are available. In the case of chemical
processes, time series techniques can be applied but the principal models
available at IH are the BIRKENES and MAGIC models. Modifications such as
the introduction of sea salt will be necessary in the case of the BIRKENES
model before application to the Loch Dee and Plynlimon catchments is
possible. Also it may be necessary to reassess the dominant equilibria
used in the model; forexample is aluminium controlled by $Al(OH)_3$,
$Al(OH)SO_4$ or by $Al(OH)_{1-x}SiO2_x$
We hope to develop a modified and combined BIRKENES and MAGIC chemistry
and couple this with the distributed models to provide an additional tool
with which to investigate catchment behaviour.

ACKNOWLEDGEMENTS

The authors are particularly grateful to the Solway River Purification
Board for providing data from the Loch Dee Study and to the British-
Scandinavian SWAP committee and the Commission of the European Communities
for providing research funding.

REFERENCES

Batterbee, R.W., R.J. Flower, A.C. Stevenson and B. Rippley. 1985. Lake
Acidification in Galloway: A Palaeoecological Test of Competing
Hypotheses. Nature V. 314 No 6009 350-352.

Beck, M.B. and Young, P.C. 1976. Systematic Identification of DO-BOD
model structure, JEED, ASCE proceedings, 102, 909-927.

Beven, K.J., Kirkby, M.J., Schofield, N. and Tagg, A.F. 1984. Testing a
physically based flood forecasting model (TOPMODEL) for three UK
catchments, J. of Hydrology, 69, 119-143.

Box, G.E.P. and Jenkins, G.M. 1970. Time Series Analysis Forecasting and
Control, Hoden Day, San Francisco.

Burns, J.C., J.S. Coy, D.J. Tervet, R. Harriman, B.R.S. Morrison and C.P. Quine. 1982. The Loch Dee Project: a Study of the Ecological Effects ofAcid Precipitation and Forest Management on an Upland Catchment in Southwest Scotland. 1. Preliminary Investigations. Fish. Mgmt. 15, 145-167.

Christophersen, N., Seip, H.M. and Wright, R.F. 1982. A model for streamwater chemistry at Birkenes, a small forested catchment in southernmost Norway, Water Resources Research, 18, 977-966.

Christophersen, N., Rustad, S., and Seip, H.M. 1984. Modelling streamwater chemistry with snowmelt, Phil. Trans. R. Soc. Lond. B. 305, 427-439.

Cosby, B.J., R.F. Wright, G.M. Hornberger, and J.N. Galloway. 1985a. Modelling the effects of acid deposition: assessment of a lumped parameter model of soil water and streamwater chemistry. Wat. Resour. Res. 21: 51-63.

Cosby, B.J., R.F. Wright, G.M. Hornberger, and J.N. Galloway. 1985b. Modelling the effects of acid deposition: estimation of long-term water quality responses in a small forested catchment. Wat. Resour. Res., Accepted.

Cosby, B.J., P.G. Whitehead and R. Neale. 1986. Modelling Long Term Changes in Stream Acidity in South-West Scotland. J. of Hydrology (in press).

Flower, R.J. and R.W. Batterbee. 1983. Diatom Evidence for Recent Acidification of Two Scottish Lochs, Nature, V. 305, No.5930, 130-133.

Gee, A.S. and Stoner, J.H. 1984. The effects of seasonal and episodic variations in water quality on the ecology of upland waters in Wales. Inst. of Water Poll. Control/Inst. of Water Eng. and Sci. joint meeting on Acid Rain, 40 pp.

Grip, H., Jansson, P.E., Jonsson, H. and Nilsson, S.I. 1985. Application of the "Birkene" Model to two Forested Catchments on the Swedish West Coast, Ecol. Bull. (in press).

Harriman, R. and Morrison, B. 1981. Forestry, Fisheries and Acid Rain in Scotland. Scott. For. 36, 89-95.

Kamari, J., Posch, M. and Kauppi, L. 1984. Development of a Model AnalysingSurface Water Acidification on a Regional Scale: Application to Individual Basins in Southern Finland. Proceedings of the Nordic H.P. Workshop, Uppsala, NHP Report No.10.

Langan, S. 1985. Atmospheric Deposition, Afforestation and Water Quality at Loch Dee, S.W. Scotland, PhD thesis, University of St. Andrews.

Mason, B.J. and Seip, H.M. 1985. The Surface Water Acidification Programme,Nature (in press).

Morris, E.M. 1980. Forecasting flood flows in the Plynlimon catchments using a deterministic, distributed mathematical model. Proc. of the Oxford Symposium, April 1980. IAHS-AISH Publ. 239, 247-255.

Neal, C. 1985. Hydrochemical Balances in the Forested Afon Hafren and Afon Hore catchments; Plynlimon, North Wales. Proceedings of Muskoka conference to appear in Water, Air and Soil Pollution journal.

Seip, H.M., Seip, R., Dillon, P.J., and de Grosbois, E., 1985. Model of sulphate concentration in a small stream in the Harp Lake catchment,Ontario. Can. J. Fish. Aquat. Sci. (in press).

Seip, H.M. and Rustad, S. 1983. Variations in Surface Water pH with changes in Sulphur Deposition. Water, Air and Soil Pollution, 21, 217-223.

Stoner, J.H., Gee, A.S. and Wade, K.R. 1984. The effects of acidification on the ecology of streams in the Upper Tywi catchment in West Wales. Envir. Poll. Ser. A., 35, 125-157.

Venn, M.W. and Day,. B. 1977. Computer aided procedure for time series analysis and identification of noisy processes (CAPTAIN) - User Manual. Inst. Hydrol., Wallingford, Rep. No. 39.

Whitehead, P.G., Neal, C. Seden-Perriton, S., Christophersen, N. and Langan,S. 1986. A time series approach to modelling stream acidity. j. of Hydrology, in press.

Whitehead, P.G. Neal, C., and Neale, R. 1986. Modelling the Effects of Hydrological changes on Stream Acidity J. of Hydrology, press.

Young, P.C. 1978. A general theory of modeling for badly defined systems. To appear in Modelling and Simulation of Land, Air and Water Resource Systems, G.C. Vansteenkiste (ed.), North Holland/American Elsevier.

Young, P.C. and Whitehead, P.G. 1977. A recursive approach to time-series analysis for multivariable systems. Int. J. Control 25, 457-482.

TESTING A SOIL-ORIENTED CHARGE BALANCE EQUILIBRIUM MODEL FOR FRESHWATER ACIDIFICATION

N. Christophersen[1], M. Johannessen[2], and R. Skaane[3]
[1]Center for Industrial Research, P.O. Box 350, Blindern, 0314 Oslo 3, Norway
[2]Norwegian Institute for Water Research, P.O. Box 333, Blindern, 0314 Oslo 3, Norway
[3]Department of Chemistry, University of Oslo, P.O. Box 1033 Blindern, 0315 Oslo 3, Norway

SUMMARY. There exist several different approaches for modelling effects of acid deposition on freshwater chemistry. All models need, however, more thorough checking against observations to improve confidence in predictions of future freshwater chemistry following changes in deposition. We use data from seven catchments in the Fyresdal-Nissedal area, southern Norway, to test assumptions in the so-called soil-oriented charge balance approach. Important processes here include ion-exchange, aluminium mineral dissolution and the CO_2-bicarbonate system. These processes form part of models like ILWAS, MAGIC, and Birkenes. Excluding samples from extended lowflow periods, theory predicts observations fairly well for streams that are not lake outflows. In the latter case the observations show more scatter.

1. INTRODUCTION

Recently, Reuss et al.(1) have given a review and critique of models for freshwater acidification. There are now several conceptually different approaches available. One group of models is denoted soil-oriented charge balance models. This group includes models by Reuss and Johnson (2), the "Birkenes model" (3), "ILWAS" (4), and "MAGIC" (5). These models more or less explicitly incorporate the mobile anion concept (6) through the charge balance principle.

Major processes include soil sulphur dynamics, cation exchange, aluminium mineral dissolution, the CO_2 - bicarbonate system, weathering, and for some models also hydrological flow paths.

For further model developement Reuss et al.(1) stress the importance of checking the models against new experimental data. Testing the models is of particular importance since many of them are being used for predictive purposes; i.e. they are used to predict changes in freshwater chemistry following changes in deposition of anthropogenic sulphur and nitrogen compounds. Testing is, however, not completely straightforward in this field. Firstly it is impossible to construct "exact" models. Any model will in practice have to concentrate on a limited number of processes. Secondly, measurements of actual freshwater and soil water composition may be of limited accuracy. This often pertains to organic anions and bicarbonate and also to aluminium chemistry at least before the early 1980ies. Both the above factors will contribute to discrepancies between observed and computed values and model assessment will necessarily become somewhat subjective.

The purpose of this paper is to test, in a fairly direct way, three key processes in the soil-oriented charge balance approach using streamwater chemistry data from seven catchments in the Fyresdal-Nissedal area in southern Norway. These processes are ion-exchange, aluminiun mineral dissolution, and the CO_2-bicarbonate system. Earlier work from one catchment in the area (7) indicated that these catchments could be suitable for such a test.

2. SITE DESCRIPTION AND METHODS

The mountaineous Fyresdal-Nissedal region lies in the western parts of Telemark county, southern Norway and experienced a dramatic decline in fish populations between 1940 and 1970 due to freshwater acidification. Except for some bigger lakes in the valleys nearly all freshwaters are now devoid of fish.

The sites treated here were established by the Norwegian SNSF-project (Acid Precipitation - Effect on Forest and Fish) in 1973 and 1974 and chosen so as to reflect differences in vegetation and soil cover in the area (8). The bedrock consists mainly of granite with darker amphibolitic intrusions in some cases. The overburden is comprised of bogs and shallow fairly organic - rich soils. Larger areas of exposed bedrock also occur. Some of the sites have spruce stands which are often poorly developed. Elevation is generally in the range 500-700 m above sea level and all catchment areas are less than 250 ha. It is important to realize that these are barren headwater catchments with shallow soils sensitive to acidification.

All sampling locations were running waters and three of the streams were lake outlets. The streams were generally sampled daily during highflow - often using automatic samplers - and weekly during lowflow from 1973 or 1974 to 1975 by the Norwegian Institute for Water Research (NIVA). From 1976 mainly weekly sampling were performed. The longest data record is for Storgama which is still sampled regularly as part of the Norwegian Monitoring Programme for Long Range Transported Air Pollutants run by the State Pollution Control Authority (SFT).

Samples were analysed for major ions including H^+, Na, K, Ca, Mg, Al, NO_3, SO_4, and Cl. Many samples taken automatically showed, however, pH values up to 0.1-0.2 units lower than those taken manually when both types of samples were analysed in the laboratory. The reason for this is still not clear and the difference is of some importance when interpreting the pH record. For aluminium total Al were measured in acidified samples and thus both organic and inorganic monomeric forms as well as some polymeric and collodial forms were included. Aluminium fractionation has, however, been carried out more recently for Storgama.

All streams are acidic (volume-weighted pH typically around 4.5) with sulphate as the major anion. Nitrate and ammonia are retained in these catchments even though the vegetation is sparse.

3. MODELLING APPROACH

The soil-oriented charge balance approach will be presented following (1). The charge balance forms the starting point which is here taken to be : (All concentrations in molar units.)

$$[H^+] + n[Al^{n+}] + 2[Ca^{2+}] + 2[Mg^{2+}] + [Na^+] + [K^+] = [SO_4^{2-}] + [NO_3^-] + [Cl^-] + [HCO_3^-] \qquad (1)$$

$[Al^{n+}]$ is a lumped representation of all positively charged aluminium species. NH_4^+ and organic anions have not been determined but are assumed negligible. Bicarbonate is generally insignificant in these freshwaters but is included here because eq.(1) will be applied to the soil solution with its higher pCO_2 (CO_2 partial pressure), and to predict water chemistry at lower sulphate depositions.

The approach requires strong acid anion concentrations (Cl^-, SO_4^{2-}, and NO_3^-) as inputs which are often taken from a submodel. We will use measured streamwater values and thus only test the assumptions determining the cations and bicarbonate. The strong acid anions in streamwater are assumed representative also for the soil solution. Quantitatively we have:

i) The carbonic acid - bicarbonate system
This is the well known relationship

$$[H^+]\ [HCO_3^-] = K_c\ pCO_2 \tag{2}$$

where pCO_2 is the partial pressure of CO_2 in soil or streamwater which are required as independent input variables. K_c is a temperature dependent constant.

ii) Cation-exchange processes
Cation-exchange equilibria are of the general form

$$K = \frac{[A^{a+}]^b}{[B^{b+}]^a} \tag{3}$$

where K is the activity ratio for two positively charged ions A and B of valence a^+ and b^+, respectively. The activity ratio depends on the fraction of each ion on the exchange complex. K is thus only a constant as long as the exchangeable fractions remain the same. This is reasonable in our case since the longest data record is less than 10 yrs. Eq.(3) is only valid in the soil solution; i.e. these equilibria are neglected when computing streamwater composition.

iii) Dissolution of aluminium
Soil - oriented charge balance models often assume equilibrium with some form of aluminium trihydroxide (gibbsite - $Al(OH)_3$). The solubility product for this reaction is

$$K_{Al} = \frac{[Al^{3+}]}{[H^+]^3} \tag{4}$$

It is noteworthy that an expression of the same form is obtained when considering H^+ - Al^{3+} exchange. Thus eq.(4) is not limited to soils where K_{Al} is in the range typical for $Al(OH)_3$ minerals (i.e. $\log K_{Al}$ = 8-10).

Given Al^{3+} aluminium speciation calculations are needed to obtain Al^{n+} in eq.(1) and close the computationel solution for the problem. Since aluminium fractionation and important complexing species like fluoride are generally lacking we have turned to a more crude procedure for estimating total Al. (See below.)

4. RESULTS AND DISCUSSION

The above model has previously been shown not to fit the observations during extended lowflow (7). This is also the case for the other six catchments. During prolonged baseflow slow weathering reactions consuming H^+ and releasing Na^+ and Ca^{2+} seem to be operating. Figures 1 and 2 show predicted and observed H^+ and Ca^{2+} concentrations as functions of the sum of strong acid anions (C_A). The values for pCO_2 (eq.(2)) in the soil was 20 times the atmospheric

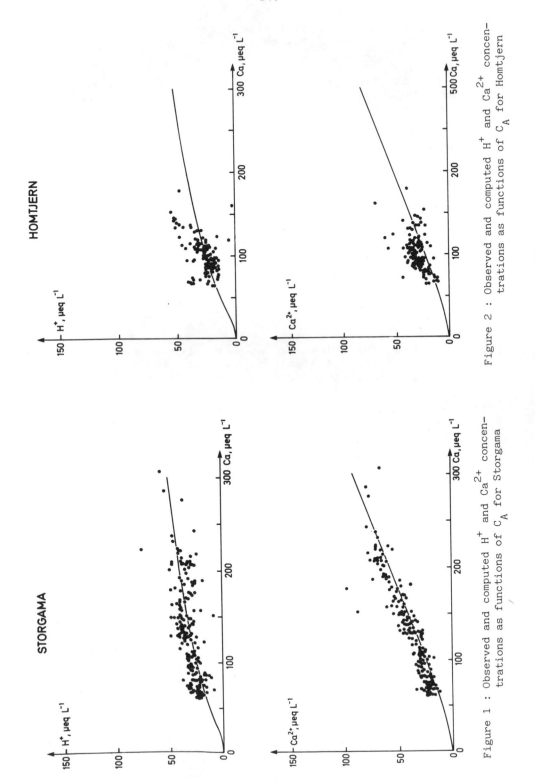

Figure 1 : Observed and computed H^+ and Ca^{2+} concentrations as functions of C_A for Storgama

Figure 2 : Observed and computed H^+ and Ca^{2+} concentrations as functions of C_A for Homtjern

pCO_2 and in streamwater two times the atmosheric partial pressure. These values were not critical. The activity ratios were determined using volume-weighted streamwater concentrations in eq.(3) and the values for log K ^1_n eq.(4) was 8.1 as used in (7). Because Al speciation could not be performed total Al was simply computed as Al^{3+} obtained from eq.(4) plus 90 ugL^{-1} to compensate for organically bound Al and inorganic complexes.

For Storgama (Fig.1) the observations fell fairly close to the theoretical lines. Homtjern (Fig.2) is a lake outlet and the scatter in the observations was considerably larger. A similar picture emerged when considering all 7 catchments of which 3 are lake outlets. For the brooks that are not draining lakes we feel the observations (excluding prolonged baseflow) fit the model well enough to support the underlying assumptions. In the lakes there were obviously other important processes going on. For instance, one might expect mixing phenomena also including incoming precipitation and meltwater to play a role. The situations where the model does not work particularly well call for some caution when applying these assumptions.

It is tempting, however, to use the results for Storgama in Figure 1 to do some tentative predictions for the case of reduced sulphate concentrations in streamwater (following reductions in deposition). If C_A declines from a present average of about 125 ueqL^{-1}, of which 80 ueqL^{-1} is sulphate, to about 85 ueqL^{-1} (corresponding to a 50% reduction in sulphate) H$^+$ is not much affected. The main response on the cation side will be in calcium. To reduce H$^+$ substantially a reduction in sulphate will have to be followed by an increased base saturation changing the activity ratios. An improvement in the base status through a net supply of base cations from weathering is presumably a long term process. It is important to realize,however, that changes in soil properties are by no means always predicted as necessary to improve freshwater chemistry. Seip et al. (9) demonstrate this for catchments with thicker soil deposits that can be described by a "two reservoir" model.

ACKNOWLEDGEMENTS

We are grateful for financial support from the Royal Norwegian Council for Scientific and Industrial Research and from the British - Scandinavian Surface Water Acidification Programme.

REFERENCES

(1) REUSS, J.O., CHRISTOPHERSEN, N., and SEIP, H.M. (1986). A critique of models for freshwater and soil acidification. Water Air Soil Pollut. (in press).

(2) REUSS, J.O. and JOHNSON, D.W. (1985). Effect of soil processes on the acidification of water by acid deposition. J. Environ. Qual., 14, 26-31.

(3) CHRISTOPHERSEN, N., SEIP, H.M, and WRIGHT, R.F. (1982). A model for streamwater chemistry at Birkenes, Norway. Water Resour. Res., 18, 977-996.

(4) GHERINI, S.A., MOK, L., HUDSON, R.J.M., DAVIES, G.F., CHEN, C.W. and GOLDSTEIN, R.A. (1985). The ILWAS model: Formulation and application. Water Air Soil Pollut., 26, 425-459.

(5) COSBY, B.J., WRIGHT, R.F., HORNBERGER, G.M, and GALLOWAY, J.N. (1985). Modelling the effects of acid deposition: Estimation of long-term water quality responses in a small forested catchment. Water Resour. Res., 21, 1591-1601.

(6) SEIP, H.M. (1980). Acidification of freshwaters - Sources and mechanisms. In Drabløs, D. and Tollan, A. (eds.) Ecological Impact of Acid Precipitation, SNSF - project, Norwegian Institute for Water Research, pp. 358-366.

(7) CHRISTOPHERSEN, N., RUSTAD, S. and SEIP, H.M. (1984). Modelling streamwater chemistry with snowmelt. Phil. Trans. R. Soc. Lond., B 305, 427-439.

(8) JOHANNESSEN, M. and JORANGER, E. (1976). Investigations of freshwater and precipitation chemistry in the Fyresdal-Nissedal area, TN 30/76, SNSF-project, Norwegian Institute for Water Research, 95pp.

(9) SEIP, H.M., CHRISTOPHERSEN, N., and RUSTAD, S. (1986). Changes in streamwater chemistry and fishery status following reduced sulphue deposition: Tentative predictions based on the "Birkenes model". (This volume)

CHANGES IN STREAMWATER CHEMISTRY AND FISHERY STATUS FOLLOWING REDUCED
SULPHUR DEPOSITION: TENTATIVE PREDICTIONS BASED ON THE "BIRKENES MODEL"

H. M. Seip, N. Christophersen and S. Rustad
Center for Industrial Research
Box 350 Blindern
0314 OSLO 3
Norway

SUMMARY. A mathematical model found to reproduce major short-term
variations in water chemistry in some streams, has been used to obtain
tentative values for pH-shifts occurring after reduced sulphur
deposition. By running the model for three years under various
assumptions about soil acidity a large number of predicted pH shifts
were calculated. The results are presented as median and 75 and 25
percentiles for 0.1 intervals of present streamwater pH. For reductions
in streamwater sulphate concentrations of 30 %, 50 % and 70 %, the
maxima for these median shifts were about 0.4, 0.7 and 1.0 pH units
respectively. The predicted shifts would greatly improve the fishery
status in southernmost Norway. The uncertainties and possible
improvements of these preliminary results are discussed.

1.   INTRODUCTION

One of the main questions related to freshwater acidification is:
What changes in freshwater chemistry and fishery status will result
from given changes in emissions of sulphur- and nitrogen compounds.
Mathematical modelling is one useful tool in attempting to answer this
question. Here we present some results obtained with the "Birkenes
model" by forcing changes in stream sulphate concentrations. The
results are highly preliminary; further calculations are in progress.
The Birkenes model has reproduced major short-term variations in
streamwater chemistry in two catchments in Norway (1,2) and one in
Canada (3, 4). The results given here are based on the version of the
model used for a tributary (Inflow # 4) to Harp lake, Ontario.

2.   THE MODEL

The hydrological model, which forms a basis for the chemical
model, consists simply of two soil reservoirs and a snow reservoir
(Figure 1). The main processes operating in the two soil reservoirs are
also given in the figure. The model is based on the mobile anion
concept (5) and stream sulphate concentrations are calculated by a
submodel. Bicarbonate and organic anions are also included. The cation
concentrations are determined by the charge balance and other equations

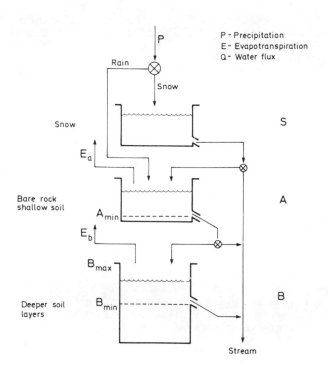

P - Precipitation
E - Evapotranspiration
Q - Water flux

Processes Operating

	Shallow soil reservoir (A)	Deeper soil reservoir (B)
H_2O	Precipitation, evapotranspiration, infiltration to lower reservoir, discharge to stream	Infiltration, evapotranspiration, discharge to stream, piston flow
SO_4^{2-}	Wet + dry deposition, adsorption/desorption, mineralization	Adsorption/desorption, reduction during warm periods
$Ca^{2+}+Mg^{2+}$	Ion-exchange	Release by weathering, adsorption/desorption
H^+	Ion exchange and equilibrium with gibbsite	Consumption by weathering, adsorption/desorption, equilibrium with gibbsite
Al^{3+}	Equilibrium with gibbsite	Equilibrium with gibbsite, adsorption/desorption
HCO_3^-	Equilibrium with a fixed CO_2-pressure	Equilibrium with a seasonally varying CO_2 pressure

Figure 1. Hydrological model used for Harp lake catchment and main processes operating.

describing the processes in Figure 1 (4). The main equations in the upper reservoir are (all concentrations in mol L^{-1}):

$$[H^+]/[M^{2+}]^{1/2} = K_{G1} \qquad (K_{G1} = 10^{-2.27}) \qquad (1)$$

$$[Na^+]/[M^{2+}]^{1/2} = K_{G2} \qquad (K_{G2} = 10^{-2.31}) \qquad (2)$$

$$[Al^{3+}][H]^{-3} = {}^*K_{so} \qquad ({}^*K_{so} = 10^{8.1}) \qquad (3)$$

$$[HCO_3^-][H^+] = K \, P_{CO_2}^{soil} \qquad (K = 10^{-7.71}) \qquad (4)$$

M^{2+} is the sum of Ca^{2+} and Mg^{2+} in soil water. $P_{CO_2}^{soil}$ is the CO_2 partial pressure in the soil. The numerical values given were used by Rustad et al. (4). In the lower reservoir we assume equations (3) and (4) to be valid, while cation exchange (eqns. (1) and (2)) is here replaced by weathering type processes. The CO_2 partial pressure in the upper reservoir was kept constant. The pressure varies seasonly in the lower reservoir but is constant during winter and snowmelt. For further details see Rustad et al. (4).

3. RESULTS

The relationship between emissions and deposition of sulphur compounds is not considered here. Neither do we discuss the time interval before a change in deposition is reflected in streamwater concentrations.
    We have assumed several scenarios which differ in concentrations of sulphate in streamwater and/or in lime potential
(LP = - log K$_{G1}$ = pH + 1/2 log [M^{2+}], cf eq 1) of the upper soil horizon.

In addition to the present sulphate concentrations (approximately 200 μeq L^{-1} during snowmelt) we have used higher and lower values obtained by changing the deposition and adjusting sulphur pools in the soil reservoirs correspondingly.
    The simulated shifts in H$^+$-peaks during snowmelt obtained by doubling or halving the streamwater sulphate concentrations are published (4) or in press (6) elsewhere.
    The simulated shifts in pH as a function of present pH levels for a 50 % reduction in sulphate concentrations in the stream are given in Fig. 2. The results are based on a three year simulation period. The lime potential in the upper soil horizon was set to 1.7, 1.9, 2.1 and 2.3 in various calculations and assumed not to be affected by acid deposition. The calculations give a scatter diagram from which the results in Fig. 2 were obtained. For pH intervals of 0.1 the median and 75 and 25 percentiles were calculated. (If the number of points was below 10, the interval was extended). Similar relationships were

computed for 30 % and 70 % reduction in streamwater sulphate concentrations. The maxima in the pH shift (interval median) were about 0.4 and 1.0 pH units respectively, compared to 0.7 for 50 % reduction.

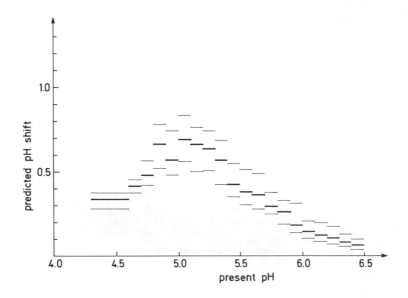

Figure 2. Predicted shifts in streamwater pH as a function of present pH level for a 50 % reduction in streamwater sulphate concentrations. Thick lines are median values, thin lines 75 % and 25 % percentiles. The figure is based on a three year simulation period using lime potentials in the upper soil of 2.3, 2.1, 1.9 and 1.7. See text for further details.

The relationship in Fig. 2 is sensitive to several assumptions, the most important ones are:

- Assumptions in the hydrological model

In the model used "piston flow" was assumed, i.e. under certain conditions water from the upper reservoir "pushes" out an equivalent amount from the lower reservoir (3). This does not change the amount of discharge, only the chemistry. The model was also tried without piston flow. This gave in general larger pH shifts. The importance of other changes in the hydrological model has not been tested, but the pH shifts are probably sensitive also to other modifications. It should be noted that the one-box model discussed by Christophersen et al. (7) normally gives smaller pH-shifts.

- Assumption about the present sulphate level

In the calculations, the sulphate level in the stream before the change was about 200 μeq L^{-1} during winter, somewhat less during the summer. The resulting pH-shifts vary somewhat with this level; lower values give smaller shifts.

- The deposition is assumed not to affect the soil lime potential

This assumption tends to make the pH-shifts too small since reduced acid deposition presumably makes the soil less acid.

- Assumptions about the CO_2 pressure

The results are sensitive to the CO_2 pressure particularly in the lower soil reservoir. In the calculations we assumed this pressure to be 10 times the partial CO_2 pressure in the air during winter and higher during summer (4).

- Nitrate and chloride have been neglected

Both in Ontario and southernmost Norway the streamwater nitrate concentrations are usually low. The neglect of chloride may be more serious in areas with high sea-salt deposition.

- The aluminium chemistry is too simple

Recent results have shown that the aluminium chemistry at Birkenes is more complex than assumed in the model. Under some conditions the Al^{3+} concentration may decrease or remain constant with decreasing pH - thus violating the gibbsite equilibrium assumption (8).

- Organic weak acids are treated in a very simple way

In the model the dissociation of these acids changes with pH, but total organic carbon (TOC) does not. There are indications that the TOC level may increase with increasing pH. The neglect of this variation tends to make the pH shifts too large.

The list is certainly not complete, and clearly the results in Fig. 2 are only tentative. In our opinion, however, the model is sufficiently realistic to make the results useful. It is probably significant that the largest pH shifts occur for present pH values around 5 with considerably smaller shifts at lower pH values.

4. IMPLICATIONS FOR FISHERY

In spite of the mentioned uncertainties it may be of interest to explore what the predicted pH shifts imply for fish populations. We therefore make the bold assumption that the relationship in Fig. 2, and those for 30 % and 70 % sulphate reductions, can be applied to lakes in southernmost Norway. A relationship between water pH and fishery status is also needed. Muniz et al. (9) have given the necessary relation for trout populations in southernmost Norway.

Combining these relations we obtain a tentative prediction of changes in fishery status in southernmost Norway for various deposition reductions. Except for a difference in the prediction of pH for a given sulphate reduction the procedure is similar to that used by Muniz et al (9). Considering the nearly 4000 lakes studied by Sevaldrud and Muniz (10), the results given in Table 1 are obtained. The improvement predicted even for a 30 % reduction in sulphate levels is quite significant; for a 70 % reduction the situation is predicted to be dramatically improved. These results are in good agreement with those presented by Wright and Henriksen (11) (with F=0.2 in their paper), though the method used is rather different.

TABLE 1. Number of lakes in the four southernmost counties in Norway with good, sparse and lost trout populations. The upper row gives observations by Sevaldrud and Muniz (9), the other rows give predictions for various reductions in sulphate concentrations in lakewater.

	Number of lakes within each category		
	Good	Sparse	Lost
Status late 1970s	899	1202	1625
30 % reduction	1426	1242	1058
50 % "	1888	1141	697
70 % "	2332	1054	340

5. FURTHER WORK

The results presented are preliminary, and further work is in progress to obtain more reliable values and explore the uncertainties. More parameters will be varied, e.g. the CO_2 partial pressure, present sulphate level, and parameters describing the hydrology and aluminium chemistry. Model parameters may be varied for two reasons, either because they differ from one catchment to the other or because we do not know the correct value. It may be convenient to consider groups of lakes (e.g. with present sulphate level within certain limits) and construct relationships between pH shift and present pH for each group.

The uncertainties in other parameters should be included by running the model with a series of values and weighting the results according to a probability distribution. Combined with the uncertainty in the relation between water pH and fishery status it should be possible to attach meaningful error estimates to the numbers in Table 1.

ACKNOWLEDGEMENTS

We are grateful for financial support from the Norwegian Council for Scientific and Industrial Research and from the British-Scandinavian Surface Water Acidification Programme.

REFERENCES

(1) CHRISTOPHERSEN, N., SEIP, H.M. and WRIGHT, R.F. (1982). A model for streamwater chemistry at Birkenes, Norway. Water Resour. Res. 18, 977-966.

(2) CHRISTOPHERSEN, N., DYMBE, L.H., JOHANNESSEN, M. and SEIP, H.M. (1983/1984). A model for sulphate in streamwater at Storgama, southern Norway. Ecol. Model. 21, 35-61.

(3) SEIP, H.M., SEIP, R., DILLON, P.J. and DE GROSBOIS, E. (1985). Model of sulphate concentration in a small stream in the Harp Lake catchment, Ontario. Can. J. Fish. Aquat. Sci. 42, 927-937.

(4) RUSTAD, S., CHRISTOPHERSEN, N., SEIP, H.M. and DILLON, P.J. (1986). Model for streamwater chemistry of a tributary to Harp Lake, Ontario. Can. J. Fish. Aquat. Sci. 43, 625-633.

(5) SEIP, H.M. (1980). Acidification of freshwater - sources and mechanisms. In: DRABLØS, D. and TOLLAN, A. (eds.), Ecological impact of acid precipitation, SNSF-project, NISK, 1432 Ås-NLH, Norway, p.358-366.

(6) SEIP, H.M., CHRISTOPHERSEN, N. and RUSTAD, S. (1986). Changes in streamwater chemistry following reduced/increased sulphur deposition predicted using the "Birkenes model", Water Air Soil Pollut. in press.

(7) CHRISTOPHERSEN, N., JOHANNESSEN, M. and SKAANE, R. (1986). Testing a soil-oriented charge balance equilibrium model for freshwater acidification, this volume.

(8) SULLIVAN, T.J., CHRISTOPHERSEN, N., MUNIZ, I.P., SEIP, H.M. and SULLIVAN, P.D. (1986). Aqueous aluminium chemistry: Response to episodic increases in discharge at Birkenes, southern Norway, Nature, in press.

(9) MUNIZ, I.P., SEIP, H.M. and SEVALDRUD, I.H. (1984). Relationship between fish populations and pH for lakes in southernmost Norway. Water Air Soil Pollut. 23, 97-113.

(10) SEVALDRUD, I.H. and MUNIZ, I.P. (1980). Sure vatn og innlandsfisket i Norge. Resultater av intervjuundersøkelsene 1974-1979 (Acid lakes and inland fisheries in Norway. Results of interview surveys 1974-1979), SNSF-project IR 77/80, NISK, 1432 Ås-NLH, Norway.

(11) WRIGHT, R.F. and HENRIKSEN, A. (1983). Restoration of Norwegian lakes by reduction in sulphur deposition. Nature, 305, 422-424.

SIMULATION OF pH, ALKALINITY AND RESIDENCE TIME IN NATURAL RIVER SYSTEMS

S. Bergström
Swedish Meteorological and Hydrological Institute

SUMMARY

Simulation of short term and long term variations in pH and alkalinity in natural river systems requires fundamentally different model approaches. These differencies are discussed. Examples of short term simulations based on the PULSE-model system are shown. A more physically based hydrological model system with a better description of the residence time of water in a basin is lined out, and examples of its performance based on simulations of records of ^{18}O are presented.

1. INTRODUCTION

The use of conceptual runoff models has so far, with few exceptions, been restricted to water quantity problems. The main reason for this is the well founded uncertainty among the modellers as concerns the physical relevance of their models. As a matter of fact, it has been shown that model structures of great diversity can produce almost identical runoff simulations. Terrestrial water quality modelling requires models with a more advanced description of the transport phenomena from rainfall or snowmelt to runoff in a river. The models must then not only be based on a sound description of the runoff generation process at a point, they also have to account for the variability of these processes in the basin.

If a model is intended only for short term variations, the geochemistry of the soil profile can be assumed to be relatively stable. This means that the variation in the water balance is a major cause of variations in hydrochemistry, which allows us to use very simplistic geohydrochemical subroutines. If, on the other hand, we wish to model long term variations as a basis for scenario simulations of the effects of changes in antropogenic emissions, we do require more advanced hydrochemical models. If these models include residence time of water in different geochemical environments as a significant parameter, we can not rely upon traditional conceptual hydrological models any longer. Instead we have to find models with more physically correct descriptions of the transit time distribution of water at various depths of the ground. This further means that we have to account for tension water in the ground as well as free drainable water.

In the following presentation, the experiences by the PULSE-model system represent a simple model for simulation of natural fluctuation of pH and alkalinity of streamflow, while

its modified successor is an attempt to arrive at a better
hydrological model, which may be a foundation for the more
advanced hydrochemical routines necessary for long term simu-
latons.

2. THE PULSE-MODEL STRUCTURE

In 1982 work started at SMHI to develop a model for terrestrial water quality problems. The background is the experience from the HBV-model for runoff simulation (Bergström and
Forsman, 1973) and its later modification for groundwater applications (Bergström and Sandberg, 1983). A detailed description of the PULSE-model is given by Bergström, Carlsson, Sandberg and Maxe (1985).

The primary objective of the work was to develop a model
for simulation of short term natural variations in pH in a
stream, caused by variable contributions from shallow acid
groundwater with short residence time and deeper groundwater of
high alkalinity. No attempt is made to model the long term
acidification of the system or the effect of variations in the
atmospheric deposition.

The hydrological snow accumulation and melt routine of the
water balance model is identical to that of the HBV-model
(Bergström, 1975), and so is the modelling of the unsaturated
zone with the only exception that capillary rise from the saturated to the unsaturated zone is considered.

The saturated zone is modelled according to the modifications introduced when applying the HBV-model to groundwater
observations (Bergström and Sandberg, 1983). This means that
there is no limiting percolation capacity within the time step
(24 hours), and quicker runoff can only occur when the groundwater level is high. In order to account for areal variabili-
ties we have introduced options for a distribution of the model
into six submodels.

In the hydrochemical part of the model no account is taken
for the acidity of the precipitation, except for the direct
precipitation on water surfaces. Instead it is assumed that the
water composition will be adjusted by cation exchange in the
humic layer, and that a considerable water exchange occurs as
the water passes through the unsaturated zone. This will level
out the effect of temporary variations in the composition of
the precipitation and also the variations in the dry deposition.

The PULSE-model uses the location or depth in the model
aquifer, from where the water drains, as an indication of alka-
linity/acidity according to Figure 1. As indicated in the figure, a seasonal component of the rate of weathering, governed
by the computed actual evaporation is included in the model, so
that the maximum alkalinity in an aquifer is $ALK_{max}(1$ in winter)
and $ALK_{max}(2)$ in summer.

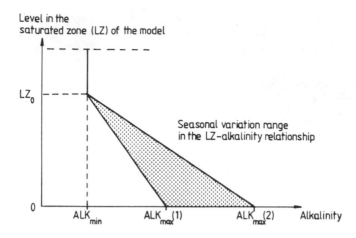

Figure 1. Principal relationship between the level (LZ) and
          alkalinity/acidity in the saturated zone of the
          PULSE-model. (From Bergström et al., 1985.)

  Together with the water balance model the subroutine of
Figure 1 will result in a general variation pattern of flow and
acidity, as illustrated in Figure 2.

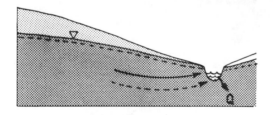

Dry period, deep ground-
water dominates, high
pH.

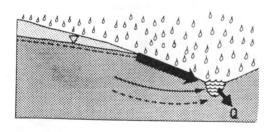

Flood, superficial
groundwater dominates,
low pH.

Figure 2. General variation pattern of flow and pH of the
          PULSE-model.

  The hydrochemical modelling in the PULSE-model is based on
alkalinity (or acidity if negative). As a final step, the alkalinity is transformed to pH according to Figure 3. This means
that we obtain simulated curves of runoff, alkalinity and pH
simultaneously.

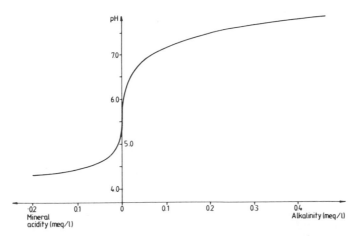

Figure 3. Principal sketch of the relation between modelled
alkalinity/acidity and pH at the outlet of a basin as
modelled by the PULSE-model. (From Bergström et al.,
1985.)

3. APPLICATION OF THE PULSE-MODEL

The hydrochemical part of the PULSE-model has four empirical coefficients, which are calibrated after the calibration of
the water balance. This means that a record of pH, alkalinity
and runoff is desired. If runoff is missing, it may anyhow be
possible to apply the model, as shown by Bergström, Brandt and
Carlsson (1985). The model is run on a daily basis with daily
totals of precipitation and daily mean air temperature as input
together with monthly standard values (30 years means) of potential evapotranspiration (Wallén, 1966).

So far the PULSE-model has been applied to several basins
in Sweden and Finland (see, for example, Bergström, Carlsson,
Sandberg and Maxe, 1985, or Bergström, Brandt and Carlsson,
1985). One example from the research basin Solmyren in northern
Sweden is shown in Figure 4.

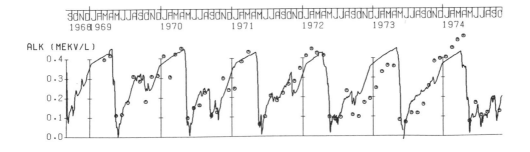

Figure 4. Example of simulation of natural variations in alkalinity in runoff by the PULSE-model. Preliminary
results from Solmyren, northern Sweden.

4. EXPERIENCES FROM THE PULSE-MODEL

The results by the PULSE-model show that a major part of the natural short term variability of pH in running water can be explained by temporary variations in the hydrological situation. The dynamics of the model are governed by simplified hydrochemical subroutines with a few coefficients which are found by model calibration for a given basin. This means that the models can be used as a reference in order to detect long term changes in the acid status of a river from short term natural variations.

Due to the simplified hydrochemical assumptions in particular in the unsaturated zone, and since no hydrochemical mass-balance is considered, the PULSE-model is not feasable for prediction of effects of changes in the acid load on a basin. It is therefore not directly comparable with the models by for example Christophersen and Wright (1981) or Chen et al. (1984). On the other hand the PULSE-model approach can be used in a large variety of rivers, due to its limited data demand.

The results of hydrochemical models of this type are very sensitive to the calibration of the water balance model. This means that the empirical hydrochemical coefficients of the model also are compensating for incomplete hydrological descriptions of the system and therefore are very difficult to generalize. This further means that at present a record on hydrochemical data is essential for parameter estimation. In the future independent assessements of the hydrochemical model parameters is an important task.

5. TOWARDS A MORE PHYSICALLY CORRECT APPROACH

The application of conceptual hydrochemical models to data on conservative tracers may provide an independent verification of the model structure and be of great value when developing the model towards a physically more sound description of run-off. Such an application was, for example, done by Christophersen et al. (1985), who attempted to simulate the flow of ^{18}O through the Birkenes basin by the Birkenes model. They concluded that the original reservoir volumes of the model had to be increased in order to obtain a sufficient damping of the input fluctuations in ^{18}O-concentration.

In an attempt to simulate records of the stable isotope ^{18}O, Lindström and Rodhe (1986) redesigned the PULSE-model and gave it a general configuration according to Figure 5.

The soil moisture storage as calculated by the original model corresponds to the amount of water held between field capacity and the wilting point. In order to describe the total volumes of water active in the hydrological cycle, an additional volume mainly representing water in the unsaturated zone below the root zone, was introduced. This volume does not affect the water balance computations.

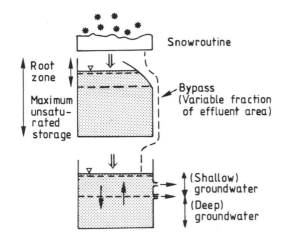

Figure 5.
Schematic flowpaths when simulating ^{18}O by the modified PULSE-model. (From Lindström and Rodhe, 1986.)

Each pulse of snowmelt or rain is individually traced through the unsaturated zone. The outflow is composed of an equal portion of each pulse in the zone. This is equivalent to what is referred to as ideal mixing (as far as transit times are concerned). Evapotranspiration is, however, assumed to take part exclusively in the root zone.

Precipitation falling on saturated areas, close to the stream, cannot infiltrate and forms saturation overland flow. This is accounted for by introducing a bypass (Figure 5).

A reservoir, representing deep groundwater and other additional water, is incorporated in the model. The volume of this reservoir is assumed to be constant and does not affect runoff calculations. Each day a portion of the water in the groundwater tank is exchanged with an equal volume from the deep groundwater storage. The fraction of deep groundwater, contributing to streamwater, will consequently rise during the recession after a flow event.

6. APPLICATION OF THE MODIFIED MODEL

Test runs with the modified model were performed for two small experimental basins in southern Sweden: Gårdsjön F 1 (0.04 km^2) and Buskbäcken (1.8 km^2). The basins represent examples of the dominating type of landscape in Sweden, i.e. coniferous forested till soils on fractured gneiss or granite rock.

While the original structure of the PULSE-model could not describe streamwater $\delta^{18}O$, sufficient damping was obtained by considering mixing with soil water in the unsaturated zone and introducing additional storage volumes, as shown in Figure 6. Final simulations for the two basins are shown in Figure 7.

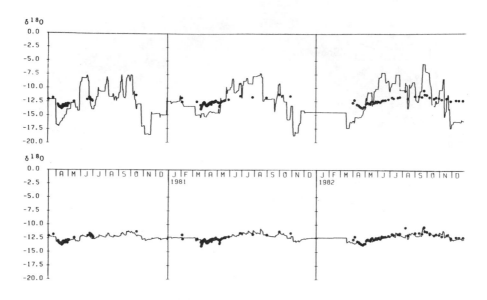

Figure 6. Simulation of streamwater $\delta^{18}O$, Buskbäcken.
(Continuous curve = simulated values,
dots = observed values.)
Above: Original PULSE-model structure.
Below: Modified model structure.
(From Lindström and Rodhe, 1986.)

7. EXPERIENCE FROM THE MODIFIED MODEL AND CONCLUDING REMARKS

The attempts to model the ^{18}O-variation pattern in small streams have clearly demonstrated that our conventional runoff models have an insufficient description of the transit time distribution of water in a basin. Although the variation pattern is better described by the modified model, there remains a lot of uncertainty as how to model the additional water storage and exchange correctly. Due to strong interaction, the simulation of ^{18}O concentrations could be obtained with variable parameter settings.

The work towards a more physically correct water balance model structure must go on along with the introduction of more advanced hydrochemical subroutines. In this process it is important that we accept and draw correct conclusions about the reality, which everyone experiences when visiting a forested till basin: The areal variability is great. The basin is not an idealized slope. It is rather like a mosaic of slopes, wet areas and exposed bedrock, which results in a diffuse and variable mixture of inflow and outflow areas.

163

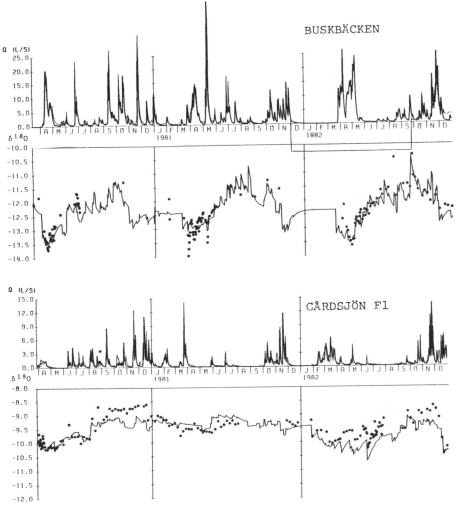

Figure 7. Simulation of runoff (temperature updated) and $\delta^{18}O$
by the modified model. Calibration periods.
(Runoff: thick curve = computed, thin curve =
observed.
$\delta^{18}O$: continuous curve = simulated values, dots =
observed values.)
(From Lindström and Rodhe, 1986.)

8. ACKNOWLEDGEMENTS
    The work reported in this paper has been financed by the
Swedish National Environment Protection Board, the Swedish
Natural Science Research Council and the SMHI.
    Thanks are due to Bengt Carlsson, who made the simulations
in the Solmyren basin, and to Göran Lindström and Allan Rodhe,
who gave the permission to use results from their recent paper
to the Nordic Hydrological Conference in August, 1986.

9. REFERENCES

Bergström, S. (1975)
The development of a snow routine for the HBV-2 model.
Nordic Hydrology, Vol. 6, No. 2

Bergström, S., and Forsman, A. (1973)
Development of a conceptual deterministic rainfall runoff
model.
Nordic Hydrology, Vol. 4, No. 3

Bergström, S., and Sandberg, G. (1983)
Simulation of groundwater response by conceptual models - Three
case studies.
Nordic Hydrology, Vol. 14, No. 2

Bergsström, B., Brandt,M. and Carlsson, B. (1985)
Hydrological and hydrochemical simulation in basins dominated
by forests and lakes.
Vatten, Vol. 41, No. 3 pp 164-171

Bergström, S., Carlsson, B., Sandberg, G. and Maxe, L. (1985)
Integrated modelling of runoff, alkalinity and pH on a daily
basis.
Nordic Hydrology 16 pp 89 - 104

Chen, C.W., Gherini, S.A., Dean, J.D., Hudson, R.J.M., and
Goldstein, R.A. (1984)
Development and calibration of the Integrated Lake-Watershed
Acidification Study Model.
In: Modeling of total acid precipitation impacts. Butterworth
Publishers: pp 175 - 203, Boston.

Christophersen, N., and Wright, R.F. (1981)
Sulphate budget and a model for sulphate concentations in
stream water at Birkenes, a small forested catchment in
southernmost Norway.
Water Resources Research, Vol. 17, pp 377 - 389

Christophersen, N., Kjaernsröd, S., and Rodhe, A. (1985.)
Preliminary evaluation of flow patterns in the Birkenes catch-
ment using natural ^{18}O as a tracer.
IHP-Workshop, Hydrological and Hydrogeochemical Mechanisms and
Model Applications to the Acidification of Ecological Systems,
Uppsala, Sept. 15 - 16, 1984. NHP-Report 10, pp 29 - 40

Lindström, G. and Rodhe, A. (1986)
Modelling of water exchange and transit times in a small till
basin.
Nordic Hydrological Conference, Reykjavik, August 1986.

Wallén, C.C. (1966)
Global solar radiation and potential evapotranspiration in
Sweden.
Sveriges Meteorologiska och Hydrologiska Institut, Meddelanden,
Serie B, nr 24, Stockholm

EVIDENCE OF STREAM ACIDIFICATION IN DENMARK AS CAUSED BY ACID DEPOSITION

A. REBSDORF and N. THYSSEN
National Agency og Environmental Protection
The Freshwater Laboratory
Lysbrogade 52, DK-8600 Silkeborg
Denmark

Summary

Some years ago a study of some oligotrophic seepage lakes in Denmark demonstrated a tendency towards acidification, which was thought mainly to be caused by the atmospheric deposition of acidifying substances.

Based on pH and alkalinity data (1977-1985) from a few small streams in two areas in Central and Western Jutland we believe that also stream waters in some of the sandy areas are in an acidifying process, which probably is due to atmospheric deposition. In one of the streams the catchment area partly consists of arable land, and here an additional cause might be ammonia fertilization of the soil.

1. INTRODUCTION

In comparison to other Scandinavian countries there is only limited information on freshwater acidification in Denmark.

Rebsdorf (1983) showed that a few oligotrophic lakes in Western Jutland were more acid in 1977-79 than earlier when compared with historical data. Thus in Grane Langsø pH had decreased from 5.6 to 5.2 from 1960 to 1977-79. The main reason for the observed decrease was thought to be atmospheric deposition of acidifying substances.

As part of a water quality surveillance programme the magnitude of acidification in two small streams (fig. 1) located on leached sandy soils were evaluated for a period of 9 years.

2. STUDY SITES

The Skærbæk is a first order stream located in an uncultivated sandy area in Central Jutland. The catchment is about 3-4 km^2. The vegetation mainly consists of grass (Deschampsia flexuosa (L.) Trin.) and heather (Calluna vulgaris (L.) Hull) and a few solitary pine trees.

The discharge and the annual variation is small in this groundwater fed stream. Discharges range around 0.03 m^3 s^{-1}.

The only submerged macrophytic vegetation is scattered patches of Potamogeton polygonifolius Pourret.

The Gryde å is a second order stream located in Western Jutland. A thorough description of the site is given by Thyssen

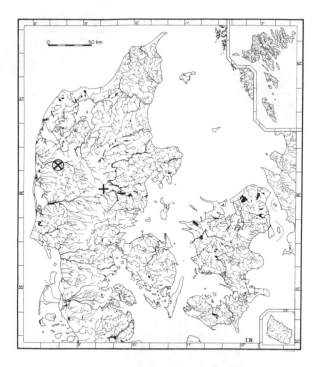

Fig. 1: Location of study sites, The Gryde ⊗ and The Skærbæk ✚.

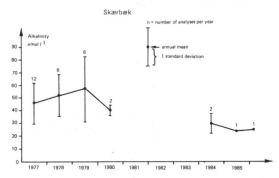

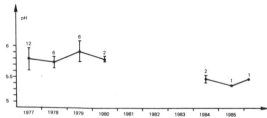

Fig. 2: Annual means of alkalinity (top) and pH (bottom) in The Skærbæk.

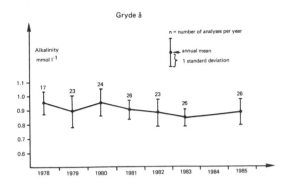

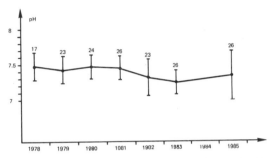

Fig. 3: Annual means of alkalinity (top) and pH (bottom) in The Gryde å.

(1982). In summary its catchment is about 15 km^2 and consists of glacial moraine. The soil is primarily sand with high porosity, infiltration and percolation, and little nutrient content. The main course of the stream lies in a military reservation with typical northwestern European Calluna heath, while upstream there is limited farming activities.

On an annual basis discharges range from 0.19 to 0.52 m^3 s^{-1}.

The submerged macrophytic vegetation show pronounced seasonal variation with dominance in spring of Ranunculus peltatus Schrank and summer to autumn dominance of Callitriche spp. and Potamogeton natans L.

3. RESULTS

In figures 2 and 3 are shown annual means of pH and alkalinity in The Skærbæk and The Gryde å, respectively.

Although the absolute value of the alkalinity in The Gryde å is rather high (mean value about 0.9 mmol l^{-1}), in comparison with The Skærbæk (mean value 0.04 mmol l^{-1}), the tendency towards a lowering of both alkalinity and pH is evident.

Based on an analysis of trends, linear regression of annual means versus time, it could be shown that pH decreased by 0.028 units per year in The Gryde å and that the observed decrease was significantly different from zero (t-test on slope) at the 95% level of significance (t_s = -2.41, df = 5). Similarly, alkalinity showed a significant decrease of 0.0114 mmol l^{-1}

year^{-1} (t_s = - 2.15, df = 5).
Since only very few measurements of pH and alkalinity were available for The Skærbæk in 1984-86 no statistical test of the observed decrease could be made.

4. DISCUSSION

The observed decreases in pH and alkalinity deserve more close monitoring and, if possible, more intensive catchment studies. The only probable cause of the observed acidification in The Skærbæk is the load of atmospheric deposition of acidifying substances as no other causes could be identified.

Also in The Gryde å the observed acidification is probably due to atmospheric deposition. However, in one of the tributaries, which partly drains farmland, some of the acidification may be due to fertilization with ammonia-containing fertilizers.

REFERENCES

REBSDORF, A. (1983). Are Danish lakes threatened by acid rain? In: Ecological effects of acid deposition. National Swedish Environment Protection Board - Report PM 1636, 287-297.
THYSSEN, N. (1982). Aspects of the oxygen dynamics of a macrophyte dominated lowland stream, pp 202-213. In: J.J. Symoens, S.S. Hooper and P. Compére (eds.). Studies on aquatic vascular plants. Royal Botanical Society of Belgium, Brussels.

CONCLUSIONS AND RECOMMENDATIONS

CONCLUSIONS AND RECOMMENDATIONS

prepared by

Arne TOLLAN, (Chairman of WPII, COST 612)
Hans HULTBERG, (Workshop-Chairman of Session I)
Alun GEE, (Workshop-Chairman of Session II)
Kari KVESETH (Representative of Norway to COST 612)
Heinrich OTT (Head of Division, CEC, DG XII/G-1)
Hartmut BARTH (Secretary of WPII, COST 612)

adopted on 11 June 1986 by the Workshop and Working Party II of COST 612

1. The workshop on reversibility of acidification demonstrated that valuable information on chemical and biological changes of aquatic ecosystems primarily due to atmospheric deposition continues to be collected and analysed in many countries, both within and outside the EEC region. Particular mention is made of recent studies of acidification of mountain lakes.

2. Data on changes in aquatic ecosystems are essential for the design of experiments and for models which predict the consequences of changes in atmospheric deposition and land use.

3. Experimental approaches (including catchment-wide addition and exclusion of acids, whole-lake acidification and restoration, and laboratory-scale manipulations), and field studies have shown that :

   - there is consensus, based on available observations and supported by model work, that aquatic ecosystems will recover from acidification once acidifying deposition is reduced ;

   - a return to pre-industrial precipitation composition would have benefits in improved water quality within a decade ;

   - land use practices can modify the effects of acidifying deposition ;

   - the important role of ammonium, including gaseous ammonia emissions, as a source of acidification (through nitrification) was demonstrated in several investigations.

4. Further experiments should aim to :

   - study the reversibility of acidification under different physiographic and hydrological conditions, particularly in mountainous areas ;

   - study the reversibility of chronic and episodic biological effects of water acidification, on individual, population, community, and ecosystem levels ;

   - further investigate the complex role of Al species, organic acids and geochemical weathering in acidified ecosystems.

5. A variety of models are in use to predict short- and long-term consequences of changing deposition patterns and quantities, including soil-water interactions. Modelling work indicates that :

- in some cases, more than 50 % reduction in total atmospheric deposition is required to prevent further acidification of susceptible water bodies ;

- although responses to reduced deposition may be rapidly evident, full recovery may be slower than the original rate of acidification.

6. Further modelling work should :

- include biological responses and should identify the key chemical, physical and biological processes and parameters ;

- improve the integration of climatological and hydrological processes in acidification models ;

- utilize available observational data for model testing in order to improve confidence in predictions, and sensitivity analysis of model assumptions.

Furthermore :

- current models should be applied regionally to catchments of different characteristics ;

- interdisciplinary cooperation between field investigators and modellers should be encouraged ;

- the potential of probabilistic models in acidification research does not yet appear to be fully appreciated.

7. The workshop recommends that working party II of cost 612 prepare during the period 1987/88 a state-of-the-art report on reversibility of acidification, to be widely circulated.

8. The workshop requests the appropriate EC and national agencies supporting research on the reversibility of acidification to note the above conclusions and recommendations.

9. Within the EEC/cost framework the reversibility and restoration of acidified aquatic ecosystems should continue to be considered. Other national and international bodies should also be encouraged to include reversibility and restoration of acidified aquatic ecosystems in their work programmes.

LIST OF PARTICIPANTS

AHL, T.,
National Environment
Protection Board
Water Quality Laboratory
Box 8005
S - 750 80 UPPSALA

AUNE, T.,
The Norwegian College
of Veterinary Medicine
P.O. Box 8146 Dep.
N - 0030 OSLO 1

BAADSVIK, K.,
Directorate of Nature Management
Tungasletta 2
N - 7000 TRONDHEIM

BARTH, H.,
Commission of
the European Communities
DG XII/G-1
Rue de la Loi 200
B - 1049 BRUSSELS

BERGGREN, E.,
Royal Norwegian Institute
for Scientific and
Industrial Research
P.O. Box 70 Tasen
N - 0801 OSLO 8

BERGSTRØM, S.,
Swedish Meteorological and
Hydrological Institute
S - 601 76 NORRKOPING

BOWMAN, J.J.,
An Foras Forbartha Teo
Pottery Road
Deans Grange
IRL - CO DUBLIN

CHRISTOPHERSEN, N.,
Center for Industrial Research
P.O. Box 350 Blindern
N - 0314 OSLO 3

COSBY, J.,
Dept. of Environmental Science
University of Virginia
Charlottesville
USA - VA 22903

ELIASSEN, A.,
Norwegian Meteorological
Institute
P.O. Box 320 Blindern
N - 0314 OSLO 3

GEE, A. S.,
Welsh Water Authority
Penyfai Lane 19
Llanelli
UK - SA15 4EL WALES

GJENGEDAL, E.,
University of Trondheim
Dept. of Chemistry
N - 7055 TRONDHEIM - DRAGVOLL

HAUHS, M.,
Norwegian Institute
for Water Research
P.O. Box 333 Blindern
N - 0314 OSLO 3

HENRIKSEN, A.
Norwegian Institute
for Water Research
P.O. Box 333 Blindern
N - 0314 OSLO 3

HIGLER, B.,
Research Institute for
Nature Management
P.O. Box 46
NL - 2956 ZR LEERSUM

HOWELLS, G.,
University of Cambridge
7 College Road
UK - LONDON SE21

HULTBERG, H.,
Swedish Environmental
Res. Institute
Box 5207
S - 40224 GØTEBORG

JOHANNESSEN, M.,
Norwegian Institute for
Water Research
P.O. Box 333 Blindern
N - 0314 OSLO

KINSMAN, D.J.J.,
Freshwater Biological
Association
The Ferry House
Ambleside
UK - CUMBRIA

KLASINC, L.,
Institute of Ruder Boskowic
P.O. Box 1016
YU - 41001 ZAGREB

KVESETH, K.,
Royal Norwegian Council
for Scientific and
Industrial Research
P.O. Box 70 Tåsen
N - 0801 OSLO 8

LANDE, A.,
Norwegian Institute for
Water Research
Groosevn. 36
N - 4890 GRIMSTAD

LENNOX, L.,
An Foras Forbartha
132, Park Ave.
Castleknock
IRL - DUBLIN 15

MOSELLO, R.
C.N.R. Ist. Italiano
Idrobiologia
L.go. Tonolli 50/52
I - 28048 VERBANIA PALLANZA

NIELSEN, H.,
Institut for Sporeplanter
Ø. Farimagsgade 2D
DK - 1353 KØBENHAVN K

OTT, H.,
Commission of the
European Communities
DG XII/G - 1
200, rue de la Loi
B - 1049 BRUSSELS

RADDUM, G.,
Zoologisk Museum
Universitetet i Bergen
N - 5000 BERGEN

REBSDORF, A.,
National Agency of
Environmental Protection
The Freshwater Laboratory
Lysbrogade 52
DK - 8600 SILKEBORG

SCHJOLDAGER, J.,
Norwegian Institute for
Air Research
P.O. Box 130
N - 2001 LILLESTRØM

SCHINDLER, D.
Fresh Water Institute
501 University Crescent
Winnipeg
CDN - MANITOBA R3T 216

SCHUURKES, R.,
Lab. of Aquatic Ecology
Toernooiveld
NL - NIJMEGEN

SCOULLOS, M.,
Dept. of Inorganic and
Environmental Chemistry
University of Athens
13a, Navarinou Street
GR - 10680 ATHENS

SEIP, H.M,
Center of Industrial Research
P.O. Box 350 Blindern
N - 0314 Oslo 3

SINDBALLE, M.,
Inst. of Plant Ecology
Københavns Universitet
Øster Farimagsgade 2D
DK - 1353 KØBENHAVN K

SODE, A.,
Department of Environmental
and Food Control
Tjørnebakken 4
DK - 6510 GRAM

STUANES, A.O.,
Norwegian Forest
Research Institute
P.O. Box 61
N - 1432 ÅS-NLH

THORESEN, B.L.,
Royal Norwegian Council
for Scientific and Industrial
Research
P.O. Box 70 Tåsen
N - 0801 OSLO 8

TIPPING, E.,
Freshwater Biological Association
The Ferry House
Ambleside
UK - CUMBRIA

TOLLAN, A.,
Norwegian Institute
for Water Research
P.O. Box 333 Blindern
N - 0314 OSLO 3

VANDERBORGHT, O.,
University of Antwerpen
SCK-CEN-Biology Dept.
Retiestwg
B - 2440 GEEL

WHITEHEAD, P.,
Institute of Hydrology
Wallingford
UK - OXON OX10 8BB

WITTERS, H.,
Belgium Nuclear Center
Baeretang. 200
B - 2400 MOL

WIUM-ANDERSEN, S.,
Freshwater-Biological Laboratory
University of Copenhagen
51 Helsingørgade
DK - 3400 HILLERØD

WRIGHT, R.,
Norwegian Institute
for Water Research
P.O. Box 333 Blindern
N - 0314 OSLO

ZOBRIST, J.,
Swiss Federal Institute
of Water Resources and
Water Control (EAWAG)
CH - 8600 DUBENDORF

INDEX OF AUTHORS

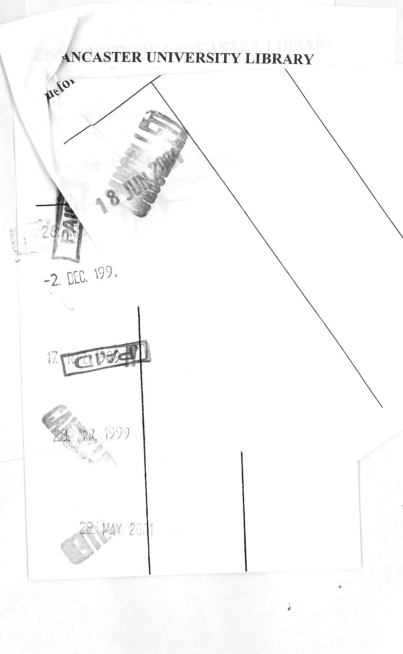